Designing Usable E

Designing Usable Electronic Text
2nd Edition

Andrew Dillon

CRC PRESS

Boca Raton   London   New York   Washington, D.C.

Library of Congress Cataloging-in-Publication Data

Dillon, Andrew.
  Designing usable electronic text / Andrew Dillon.—2nd. ed.
     p. cm.
  Includes bibliographical references and index.
    ISBN 0-415-24059-X
    1. Text processing (Computer science) 2. Electronic publishing. 3.
  Human-computer interaction. I. Title.
  QA76.9.T48D55 2003
  005.7—dc21   2003004075

Visit the CRC Press Web site at www.crcpress.com

© 2004 by CRC Press LLC

No claim to original U.S. Government works
International Standard Book Number 0-415-24059-X
Library of Congress Card Number 2003004075

Contents

Acknowledgements

"To err is human, to forgive, design"

Andrew Dillon (1992)

The first edition of this book owed so much to my working experiences at the HUSAT Research Institute of Loughborough University in England, where I spent many happy times between 1986 and 1993. Long before the current vogue of marketing user-centered methods, HUSAT advocated usability and an appreciation of the users' and organisations' perspective in developing acceptable new technologies. More than this, HUSAT convinced the design and manufacturing worlds to support its work.

Since the first edition was published I moved to the USA and became a faculty member first at Indiana University, where I found myself crossing so many disciplinary divides that I have ended up with official listings in five units (information science, informatics, cognitive science, computer science and instructional systems technology). These multiple identities speak volumes both for the academic open-mindedness of Indiana University and for the fuzzy boundaries that exist when one starts exploring the human response to information technology. At Indiana I worked with some excellent faculty and students, and it is the latter group who have most shaped my ideas in the intervening periods between these editions (not least since many have had to read and comment on this book and my various other writings as part of their class activities). To name them might be dangerous but I am particularly grateful to the following: Misha Vaughn (now a senior usability specialist at Oracle Corp.), Mike Morris (now a faculty member at the McIntyre School of Commerce at the University of Virginia) and Michael Chui (now with Mckinsey and Company, Consultants).

Since leaving Indiana to become Dean of the School of Information at The University of Texas at Austin I have had the opportunity to work from the ground up in planning a new kind of school to deal with the human side of information. At Texas I must thank Kayla Chioco for her graphic and proofreading skills, not to mention her general good cheer when faced with the exploding manuscript.

This edition was delayed longer than one might have expected so I must thank Sarah Kramer at Taylor & Francis for her patience with the inevitable slipping in my schedule. Patience is also one virtue my lovely wife Marian has had to perfect in dealing with me. My excuses are genuine – a wonderful son, Oscar, came into the world and threw all my plans out the window. But such is life and the extended period has, I hope, improved the resulting text considerably. Oscar is born in the digital age but shows all the early signs of being both a computer and a book lover. What joy awaits him!

Andrew Dillon
School of Information
February 2003

1 Reading, books and electronic text

In a decade or so, the book as we know it will be obsolete
Dave Jonassen, 1982

Introduction

In a 1994 interview with Parade Magazine, Carl Sagan, the noted astronomer, said, 'books tap the wisdom of our species, the greatest minds, the best teachers from all over the world and from all our history. And they're patient'. I can think of no more accurate description of the ideal of the textual artifact. But books are also big business. In 2001 it is estimated that more than 1.6 billion books were sold in the US, generating $25 billion of revenue. Electronic books sell in minute figures in comparison, but even they are experiencing double digit rates of growth in recent years as new technologies evolve. While it is typical to discuss such developments in terms of falling hardware costs and technological advances, it is, at least for the humanists and social scientists among us, more interesting to observe such developments in terms of their influence on human activities. While theorists talk of the information age, and now even the postinformation age, we are in practice creating an information world where microprocessors interface between innumerable, as well as previously unimaginable, activities and us.

The present book is concerned with one such development, the use of information technology to support the activity known as 'reading' and, in so doing, to challenge the supremacy of paper as the most suitable medium of text presentation. This area is receiving a lot of attention as hypermedia and the World Wide Web (which I shall from now on refer to generally as 'the Web') give new substance to old ideas, but such attention is typically directed more at developing the technology than considering how and why it might be useful. This book is concerned directly with the reader and, thereby, the text designer, author and publisher who must cater for her. In so being, it touches on numerous related areas where documents of any kind play a part in human activities: education, training, leisure, work and so forth. However, it is the user or reader and the design of an information technology that suits her[1] which is my focus here.

The emergence of electronic text

For a medium that is so new it is perhaps surprising that a history of electronic text can be even considered, never mind described. While the current advertising hype is that e-books represent the most important revolution since the Gutenberg press, the idea of using the electronic medium to support reading is less revolutionary than evolutionary in that it can be traced back more than 50 years now and no self-respecting writer on the subject of hypertext ever fails to mention such visionary thinkers as Bush (1945), Engelbart (1963) or Nelson (1987),[2] who in their own way advanced (and in some cases continue to do so) the concept of instant access to a world of knowledge through information technology. These thinkers paved the intellectual path to hypertext[3] and the Web with its underlying philosophy that humans should be able – from their desktop, laptop or palmtop; at work, at home, or on the move – to locate, retrieve and use easily the store of human knowledge that lies in books, journals and associated materials in libraries the world over.

Despite its ancestry or philosophy, electronic text has had to wait for a technology to develop before such fantastic ideas could be embodied. The computer is that technology and only comparatively recent developments in microelectronics have enabled the concept of electronic text to be seen and not just heard. Feldman (1990) points out that despite the advocates of previous decades, it is only in the 1980s that electronic text could really be said to have arrived. Prior to that it was conceived, talked about, and its potential imagined, but it did not truly exist.

Spring (1991) puts it nicely when he states that:

---

1   Throughout the text, footnotes are used to examine related points or inform interested readers on alternatives and caveats. The fact that you are reading this one has probably demonstrated to you that they are not the most usable literary devices, particularly where publishers make authors place them off the page of occurrence. While I shall keep footnotes to a minimum in the book, I encourage you to read them as you come across them for maximum appreciation. In so doing, you might become acutely aware of one possible advantage for electronic text.

2   The comparative recency of this reference stems from the fact that much of Nelson's work is self-published and/or distributed. His ideas actually gained currency far earlier than this date suggests.

3   Hypertext is an electronic form of text presentation that supports the linking of nodes or chunks of text in any order. It has been defined and explained so often in the previous 10 years that I will not attempt to explain it further. If any readers find this description insufficient or really do not know what the term implies, a quick read of any one of a dozen introductory texts on the subject will provide the answer (see for example, McKnight *et al.* 1991).

During the 1980s there was a little-noticed change in the world of computing. One day during that decade, I guess it was about 1985, more computing cycles were devoted to manipulating words than were devoted to manipulating numbers.

(p. iii)

The precision of the timeline is debatable but it is not difficult to see why the personal computer boom of the 1980s coupled with developments in digital information storage and presentation have made electronic text both feasible and culturally acceptable. Both these aspects are necessary for electronic documents to succeed. It is not enough that it can now be done, that electronic text can, for example, reduce the 20-volume Grolier Encyclopedia to a single compact disc (and still leave more than half of the disc free), or that one can purchase a 100-book collection of great literature on disc for less than $15, or even that a university campus can declare its intention to become paper-free by Fall 2003;[4] the world needs to be ready for electronic text. Readers must appreciate its relevance, its potential advantages and, more importantly, they must want to use electronic text if it is to succeed. We are not quite there yet.

While the information culture is emerging, the acceptance of electronic text continues to lag behind its technical feasibility. As with most, if not all, technical innovations, the claims of its advocates wildly surpass the reality of its impact. With the exception of a small number of researchers, designers and keen amateurs, the idea of reading lengthy texts in their electronic as opposed to paper form tends to be viewed negatively rather than embraced wholeheartedly. In no small way that is why you are probably reading this sentence through the well-established medium of paper.

It will take time and effort to identify the optimum forms for electronic text. Currently, designers lack the expertise and experiences that have evolved with paper text and the present work seeks to contribute to the increased research and development effort on this front. These are early days for the new medium (even if it is possible to distinguish between generations of electronic text) and I wish to make my position clear at the outset: one should avoid seeing electronic text as a competitor to paper in some form of 'either–or' challenge for supremacy. It is not inevitable that electronic text will replace paper as some early writers have suggested, but it might displace it as more and more human activities become mediated by information technology. This should not be allowed to happen by accident though, as a side-effect of increased computerisation; we must seek to actively influence the process for the better so that the positive aspects of electronic text

---

4 This gem of an idea was launched by the West Des Moines campus of Des Moines Area Community College in Fall 2002. 'Hopefully within a year', said the Dean, 'we'll have the whole campus paperless' (see *Wired News: Who Needs Paper? Not Iowa College* http://www.wired.com/news/school/0.1383.53747.00.html).

are accentuated. To achieve this, psychologists, information scientists, sociologists, writers and (most importantly) readers must influence the design process.

The history of electronic text, and of information technology in general, is still being written and will be shaped by many forces. There will be many unintended consequences of its use, new and unforeseen uses, and resistance to it in many quarters. All well and good – we should never accept a technology at face value but should strive to test it, argue about it, modify and shape it until it suits our needs as stakeholders and users, no matter how benign the original design intent.

The aims of the book

The major aim of the present work is to examine and subsequently describe the reading process from a perspective that sheds light on the potential for information technology to support that process. Current research suggests that paper is by far the preferred medium for reading, though there is less consensus on why this is, in fact, the case. It is clear that simply transferring paper formats to the electronic medium is insufficient and often detrimental to use. Therefore, clarification of the future role for electronic text in our information world would seem to be an issue worthy of investigation. Tackling it as a social scientist, the author is far less concerned with technical feasibility, that is, Can it be built? (although of necessity at times this question rears its head) than with how human cognition and behaviour place constraints on, or provide clues to, the usability (and hence the acceptability) of current and future technologies.

This, however, is not a social science text, and it most certainly is not a text on the psychology of reading as that descriptor is usually understood. But it is a book about reading and the examination of that process from a primarily psychological and social informatics perspective. If it needs classification it could be described as a user experience or human factors[5] book but even then it may not conform to others' expectations of that subject's style. If classification proves awkward then so much the worse for the classification. The design of usable electronic text and the study of human information usage are issues that deserve direct attention and, as is discussed frequently throughout this book, rigid disciplinary boundaries are not helpful here.

Traditionally, ergonomics (or human factors) has offered itself to designers and engineers as an evaluative discipline, equipped with the tools and

---

5 Ergonomics is the scientific study of the person in relation to the designed environment (usually work). In the USA, ergonomics was more frequently known as 'human factors' until recently, when a move towards using the more European term 'ergonomics' has been observed. The terms will be used interchangeably throughout this book, along with a third term, user experience.

methods to assess the performance of human operators with developed systems. In recent years, as a result of the need for more rapid design cycles and increased competition among developers, a need for earlier inputs to the product life cycle has arisen. Such inputs, in the form of models, guidelines, checklists and design tools, attempt to package human factors knowledge in a form suitable for engineers to consume and apply. This has not proved an easy task and there are many in the human factors discipline uncomfortable with this role (not least the present author).

This book does not concentrate on the issue of technology transfer between ergonomics and design but is aware of its existence as a yardstick against which, rightly or wrongly, the value of current human factors work is often measured. Consequently, such issues cannot be avoided in contemporary ergonomics, and a second aim of this book is to develop a framework for considering user issues that is potentially applicable to the earliest stages of electronic text design. However, it should be emphasised here and now, it is not expected that such a framework could just be handed over to designers and programmers and flawlessly applied by them as so often appears to be the intention of user experience professionals. The emphasis throughout the work is therefore less on empirical investigations of various user interface variables (though these are to some extent present) and more on identifying the primary human factors underlying document usage through knowledge elicitation techniques and observation of usage patterns, with a view to forming these into a conceptual framework that can guide discussions, musings and evaluations of design ideas.

The scope of the book

In simple terms, this work is concerned with the human as reader, that is, user of textual information. However, its remit is broad by comparison to much of the theoretical psychological work in this area, which tends to define reading narrowly as the transformation of visual images to perceived words or the extraction of meaning from structured prose. Such positions, though extremist, are both tenable and frequently published. Indeed much of the work in experimental psychology on reading assumes one or other interpretation (see for example, Just and Carpenter's 1980 model of reading, which includes comprehension, and Crowder and Wagner's 1992 description of where reading begins and ends, which explicitly excludes it). Instead, this book covers the range of issues involved in using such material, from identifying a need for it, locating and selecting it, manipulating it and ultimately processing it. Therefore, while its interests are primarily psychological, the consideration of alternative perspectives from disciplines such as information science, computer science, education, sociology, literary theory and typography is both necessary and insightful.

In the present context, therefore, 'reading' implies situations where the human will engage the medium to perform any of the range of activities typically afforded this descriptor in common parlance. Thus it covers a variety of scenarios ranging from proofreading a document to examining the contents of a book to see if it contains anything of interest, but omits those that have reading as a component but necessarily secondary part, such as document filing. Furthermore, included under this term are the associated activities of location and manipulation of textual information that invariably precede and are concurrent with these tasks in the real world. How far one extends this is a matter of common sense. Obviously, walking into a library or bookshop, or logging on to one's internet account necessarily precedes the act of reading a text there but should not be considered part of the act of reading itself. However, within the broad reading-task scenario, searching for a book or browsing the spines of numerous journals to locate a specific edition is part of reading in the sense used here.

By text is meant any document, with or without graphics, which can be presented to a reader as an information source. Thus it includes those documents that we are typically exposed to in everyday life such as newspapers, books, email, magazines, technical manuals, e-commerce Web sites and so forth. Though termed text, this descriptor might include those documents that have a large graphical content (such as catalogues) but not those that are primarily graphical such that they relegate alphanumeric text strings to secondary importance (for example, maps). That said, much of what follows in terms of usage and structure of information would seem relevant to the design of even highly graphical information displays.

The terms 'electronic text' and 'digital document' are used by means of contrast with paper documentation, that is, they mean any text that is presented on a computer screen. For the purposes of this book the terms electronic and screen-presented text are used synonymously and imply presentation via computer screens, be they on handheld devices or large monitors. They do not refer to any other form of screen-presented text such as microfiche, microfilm or slides that involve magnification and projection rather than electronic processing. The term includes hypertext and non-hypertext. Like its paper equivalent, electronic text may contain graphics. Generically, the term information space is used to cover all published materials: text, hypertext and hypermedia. Where it is employed in this book its meaning is implied by the context of use unless otherwise stated.

Obviously it is impossible in the present situation to cover all manifestations of reading texts as the terms are defined here, and indeed the book concentrates primarily (though not exclusively) on scientific literature such as academic journal articles or technical manuals for its empirical investigations. However, even academic articles are lengthy texts that, it will be shown, are read in a variety of ways that extend their comparability with other document forms. Similarly, the use of software manuals, booklets and word processor documents also reported here broaden the coverage of the work. Thus the issues raised and concepts presented in the final framework are intended to be generic and applicable to most text forms and reading situations in as much as electronic media might influence their interaction.

A note on methods

This work, by choice, avoids many of the issues of learning to use innovative technology which some would see as a natural role for a human factors researcher (and the subject matter of numerous workers who claim to be examining usability when they are really evaluating learnability, which is not the same thing). It is not that such research is seen as irrelevant but that the author believes that well-designed systems should start from a premise of supporting specific users performing certain tasks rather than worrying prematurely about ease of learning. In this application domain, design is necessarily speculative, there are few, if any, rules or established systems to react to or design against. Contrast this, for example, with designing a new text editor where not only does a large body of knowledge on how users perform such tasks exist, but designers can also examine numerous existing products to inform their own design (which may account for the preponderance of comparisons with paper one finds in the e-book world). Consequently, the author sees the role of psychology and other social sciences in this area as a dual one of guidance and suggestion, using knowledge of human cognition and performance to constrain the number of potential design options while informing speculation on how humans might like things to be. Such work necessarily precedes learnability research.

The stated aims and approach of this work have dictated the methods employed, not only in this book but also in my professional work. This text is an applied work, a study of human factors efforts carried out during the development and evaluation of real products. In order to identify how electronic text systems are designed and the best role for human factors knowledge in this process it is necessary to involve oneself in the process, to be part of a design team, to develop electronic texts and assess the consequences of one's work. Only in this way can one really appreciate what is needed, what questions arise, what type of human factors input is useful and what are the limitations of the discipline's (and one's own) knowledge. Theorising from without may have proved intellectually stimulating but would have been insufficient. To paraphrase Card *et al.* (1983), 'design is where the action is' in human–computer interaction (HCI).

On the face of it, the human scientist would appear not particularly well armed for action of this kind. The traditional strengths of psychology and ergonomics lie in designing and conducting formal experiments, planning work in detail before carrying it out, controlling for all undesirable sources of variance, and reporting the results in a conventional academic form. As a result of such an approach, a substantial literature has emerged on the presumed usability of various interface features or the significant problems associated with certain products. Essential as such work is in building up the bedrock of empirical knowledge, on its own it cannot provide the answers to many of the questions posed here.

Examining the issue more deeply, however, one might come to appreciate that the human scientist is the ideal person to become involved in the design of interactive products. Equipped with knowledge of human behaviour and dispositions, skilled in the consideration of how certain design features influence performance, the human scientist can make the distinction between popular conceptions of users based on opinion and myth, and accurate models based on reasoned argument and psychological findings. One should be able to distinguish between occasions when approximate answers will suffice and when only formal experimental evaluation will provide answers. Most importantly, one should be able to identify gaps in the knowledge base of design that only the human sciences can fill or hope to fill. In short, the human scientist may be seen as the only suitable candidate for the job.

This is the philosophy of the present work. Involvement has been achieved over the years by working on designs and research in the UK and the USA. As a new researcher to the field I joined the HUSAT Research Institute in Loughborough University,[6] the largest university-based research and consultancy institute in Europe dedicated to the application of the tools and techniques of the human sciences in technology design. There I became a member of a research team (with Cliff McKnight and John Richardson) specifically investigating electronic text design. In the subsequent period this team worked on four long-term research projects related to electronic documentation and human information usage. These projects were coupled with a variety of short-term consultancies on human factors for numerous industrial software companies and departments throughout Europe. Our philosophy was always one of 'show me the data', and we published many studies and critiques of designs in those years based on our user tests.

In 1993 I left HUSAT for the US, ending up in my present position in Austin at the new School of Information at The University of Texas, after 8 years at Indiana University where I directed the Usability Laboratory at the School of Library and Information Science and developed a Masters Degree in HCI through their new School of Informatics. A sabbatical from Indiana in 2000–2001 enabled me to put the revisions together for this second edition. While at Indiana I also revamped the original framework into the version (called TIME) that is presented here. Again, the testing ground was the realm of real-world products but this time I was assisted in testing the framework by numerous students who ran with it in my advanced usability courses. There are too many to name here but I am grateful to all of them and am

---

6   HUSAT stands for Human Sciences and Advanced Technology. Founded in 1970, this group has had a profound impact on technology design and the thinking that social science has relevance for engineering practice.

rewarded to see so many of them now successfully employed as HCI professionals throughout the world.

The nature of the work and the impact of the findings on real-world applications mean that most of the studies reported here were not laboratory exercises isolated from practical concerns but investigations carried out during design processes to provide answers to genuine questions, to resolve design issues or to test specific design instantiations. It is an example of the psychologist as applied scientist, part designer, part team member and part user, all the while monitoring his and his colleagues' own work in a meta-analytic fashion. It is my contention that such a process is not only a worthwhile method of research but also the only sure way for suitable knowledge of this field to be gained. Interested methodologists who would like more information on the use of research in design might find Klein and Eason (1991) a useful source.

Generally, the techniques and methods employed in this book vary from the experimental to the exploratory. The use of a method was determined by the type of information sought – what needed to be known led to the choice of investigative methodology. Expertise in or familiarity with a technique was never considered sufficient justification for its employment. For example, at the outset it became clear that information on how readers view texts and their interactions with them was in short supply in the literature. This gap in the knowledge base is in part due to the inherent difficulties in capturing such information in a valid form. Experimental techniques are impracticable in such situations and reliable questionnaires on such matters do not yet exist. In order to overcome this, information was gathered employing a mix of knowledge elicitation techniques from the 'harder' (in the comparative sense only) or more objective such as repertory grid analysis, to the 'softer' or more subjective ones such as interviewing until satisfactory answers were obtained. In this I took inspiration from Binder (1964), a psychologist and statistician who wrote:

> We must use all available weapons of attack, face our problems realistically and not retreat to the land of fashionable sterility, learn to sweat over our data with an admixture of judgment and intuitive rumination, and accept the usefulness of particular data even when the level of analysis for them is markedly below that for other data in the empirical area.
>
> (p. 294)

I have provided the sweat and, I hope, the correct 'admixture of judgment and intuitive rumination,' and it is for the reader alone to determine whether the results have achieved the appropriate status of usefulness.

Outline of the book

The book commences with an examination of the concept of usability as it is applied to the design of interactive technologies and the potential application of human factors or ergonomic knowledge and techniques in this domain. Following this is a thorough review of the experimental literature on reading from paper and screens. This review is divided into three major parts. The first describes the reported differences between reading from paper and from screens. The second part reviews the analyses of these differences that have resulted, which can be seen as an attempt to identify the basic differences between the media and subsequently isolate the crucial variables in terms of four levels: perceptual, motor, cognitive and social issues. The third part of the literature review highlights potential short-comings in the comparative feature-based approach and raises issues relating to text and task variables.

Chapter 4 concentrates on the value of the existing human factors findings to electronic document designers, highlighting the underlying acceptance of unidisciplinary (primarily cognitive) views of the reading process manifest in the literature. The problems of applying findings from this work to the design and evaluation of an electronic text system are highlighted by reference to a case study carried out by the author.

In order to overcome some of the perceived shortcomings of previous work, particularly in relation to reading context, an analysis of the cate-gorisation of texts, including readers' views of text classifications, is presented in Chapters 5 and 6. This work emphasises the world of texts as partly given, partly constructed by the reader, and acts as a stimulus to more detailed methods we can use to observe readers. Chapter 7 considers further the literature on readers' impressions of structure and shape in information space, the limitations of the navigation metaphor, and the emergence of digital genres.

The TIME framework proposed in Chapter 8 represents the human factors involved in using a text and suggests the variables to consider in designing an electronic document. The framework consists of four interactive elements that reflect the issues dominating the reader's attention at various stages of the reading process. The interactions between these elements are described and the role of such frameworks in human factors work is assessed. Chapter 9 explores examples of the TIME framework in action, using it to derive experimental hypotheses about users, to account for verbal utterances of users, and showing how the framework can be used to structure quick usability walkthroughs of e-document designs.

The final chapter highlights areas for future research. A sequence of human factors inputs to the design stage, which should aid usability, is made explicit in this chapter. The final section of the book assesses the realistic prospects for electronic text in the information world that is emerging.

Unlike most digital document advocates I will not encourage you to jump all over the book – I have written it this way in order to highlight how the ideas developed and are related. However, as experimental evidence has demonstrated repeatedly to me that educated readers rarely read a book in a completely linear fashion anyway, I should neither encourage you (as you need no such encouragement) nor discourage you (because it is likely you are reading this section too late for it to make a difference!).

2 Electronic documents as usable artifacts

Sirs, I have tested your machine. It adds a new terror to life and makes death a long felt want.

Sir Herbert Beerbohm Tree, on examining a gramophone player

The emergence of usability as a design issue

Electronic documents are one more application of an information technology that has proved immensely successful in supporting tasks as diverse as flying an aircraft and handling financial transactions. The computer has expanded beyond its initial remit as a number cruncher for scientists and become a storer, manipulator and presenter (and even a supposedly 'intelligent' analyser) of all kinds of information. What really could be more appropriate for a so-called 'information technology' than to store, manipulate and present text for readers?

Before examining issues of direct concern to electronic documents, it is worth considering their emergence from the broader technological perspective that became known as the 'information revolution'. Without wishing to delve deeply into the history of computing technology, it is important to appreciate that computers moved from expensive, large and dedicated systems used only by specialists in scientific or business domains in the 1950s and 1960s, through a series of stages involving ever smaller and cheaper instantiations, operated by less specialist users, employing a variety of higher level languages in the 1970s. The shift from the bulky mainframe to the desktop computer late in that decade paved the way for widespread usage by people not particularly well versed in programming languages, operating systems and hardware issues.

As technological developments in the computing field advanced, it became clear that cost and functionality were no longer the major obstacles to massive uptake in society. Falling comparative hardware and software costs meant that computing technology became affordable for many applications. However, bottlenecks in the technology's acceptance appeared largely as a result of problems associated with users of the new systems.

Originally, the users of this technology were experts in its design and inner workings, often building and maintaining it themselves and certainly trained and knowledgeable in the effective operation of their computers. Typically, this meant spending several years learning to write and read specialised programming languages in order to 'instruct' the machine. However, as early as the late 1960s some computer scientists[1] (as they became known) noted large variance in the time it took programmers to produce the same codes and that different end-users responded differently (and unpredictably) to the finished design (see for example, Weinberg 1971). It started to become clear that human costs were also important in the development and particularly the usage of this revolutionary technology. As the new waves of computing applications emerged, it was human resistance that came to be seen as holding up the new technology's exploitation.

Resistance is perhaps too strong a word for this phenomenon as it implies active, Luddite countermeasures to spoil the advance of information technology. More accurately, the cause of delays in uptake lay more with flaws in design than in any widespread movement against the digital revolution. For any technology to reach a wide audience of potential users it must minimise the amount of specialist knowledge or training required to make use of it, otherwise it is destined to appeal only to trained or special-ist user groups. The rise of the 'casual user', persons with little or no specialist training or knowledge (noted informally by many early in the 1970s but formally named first by Cuff 1980), was an obstacle to uptake that could only be crossed if the technology was made suitably easy to use by its developers.

As a result, in the 1970s a research effort started in earnest that carries on to this day. This research took as its subject matter the process of human–computer interaction with a view to identifying how computing artifacts could be designed to satisfy functional requirements while min-imising users' difficulties. The field became known by the same name as its subject matter – human–computer interaction or HCI[2] – but more accurately should be described as an extension or application domain of the existing discipline of ergonomics or human factors. In fact, ergonomic concern with computer interfaces can be traced as far back as the work of Shackel (1959), considered by many to be the first published paper on HCI.

HCI focused attention on the interface between the person and the machine such that interface design has now become a central concern to interactive product developers. The interface is essentially a communication channel that affords the transfer of modality-independent information

---

1 Of course, it must be noted that 'computer science' has become so established that many have forgotten the quip that any subject with the term 'science' in its name cannot be one!

2 In the USA this acronym was often shortened to CHI (computer–human interaction), a reversal that some human scientists are quick to point to as an example of technologists putting the computer before the human!

between the user and the computer. Possessing both physical (for example, keyboard, mouse, tablet, microphone, screen, etc.) and representational qualities (for example, it is designed to look like a desktop, a spreadsheet or even a book, etc.), the user interface is, for many people, all they see and know about the computer. Immense research effort has been expended on studying how this transfer of information can be best supported so users can exploit the undoubted power of the computer to perform tasks. Certainly, progress has been made and contemporary interfaces are often (but not always) far superior to typical user interfaces of even a decade ago. Yet problems still exist and user error, resistance and dissatisfaction often still result from poorly designed technology. The field of HCI still has much work to do to ensure user issues are adequately addressed in design.

This brief introduction summarises significant developments in recent design, technological, industrial and social history that deserve entire books to themselves to cover appropriately (see for example, Forester 1985, Eason 1988 or Borgman 2000). However, it is against this backdrop that we must examine the central issue in the design of electronic documents – their usability.

Usability as part of the product acceptability equation

The success of any interactive product or system is ultimately dependent on it providing, at an appropriate price, the right facilities for the task at hand in such a way that they can be effectively and satisfactorily exploited by the intended users. If it can satisfy these criteria then it can aptly be described as acceptable.

Shackel (1991) discusses product acceptability as an equation involving the relationship between functionality, usability and cost. Functionality refers to the complete range of facilities offered by a tool or product. Basically, any technology offers its users the potential to perform a variety of actions. Some products offer more functions than others and in certain product domains extra functionality is often used as a major selling point in marketing the technology. Obviously, for a product to be acceptable it must offer users the range of functions they need.

However, merely providing the facilities to perform a task serves little purpose on its own if the user cannot discover them, fails to employ them effectively through error or misunderstanding or dislikes them sufficiently to avoid usage. Such problems are not *functionality* issues but *usability* issues and it is these that are determined largely by the user interface. It is not enough that a system is efficient, cheap and highly functional in task terms. If users have problems using it then such features are never exploited and the product must be deemed at least a partial failure.

This point can easily be appreciated by considering the typical office telephone. Many phone sets offer the potential to transfer calls to colleagues, to form conference calls, to answer a phone on another desk that is linked in a network, or to place call-back commands on engaged lines. These are all *functions*. However, many users are incapable of exploiting these functions for a variety of reasons, for example, they were never trained in using the phone, they cannot remember the codes required, they do not want to read the accompanying manual (if they can still find it). The keypad on the phone itself rarely affords sufficient clues and a user is therefore required to remember, for example, that **3 might answer a colleague's ringing phone. These issues are a reflection of the interface design of the telephone set and therefore provide a statement of its *usability*. Such a design clearly reflects an emphasis on functionality at the expense of usability, offering multiple options via a constrained interface (a keypad) and implementing them in a manner divorced from the capabilities, cognitive tendencies and motivations of most users.

Usability is an increasingly popular concept that is at the heart of user-centred system design, a philosophy of design that will be outlined later in this chapter and recurs as a theme throughout this book. Varyingly equated with such concepts as 'user-friendliness'[3] or 'ease of use' (see Stevens 1983 for an early analysis of the vague underpinnings of this concept), usability is often employed without formal definition. Indeed, in the literature on HCI one can find at least three distinct bases for definition:

- Semantics – aimed at explaining the meaning of the word in common-sense terms, for example, usability means 'user-friendly', 'easy-to-use' or 'transparent'.
- Features – aimed at providing a list of attributes an interface should possess to be usable, for example, windows, mouse, menus, consistent labels, etc.
- Operational – aimed at defining usability through measurable attributes of interaction.

Of the three, feature-based definitions have permeated popular culture to the point that usability is frequently equated with the provision of a graphical user interface (GUI) or WIMP[4]-style interaction. Conformance with standard Windows or MAC interface forms is considered sufficient in these terms to be usable. However, such a view is inaccurate and misleading. There are no standard sets of features that work for all users in all situations. Furthermore, the feature-based approach locks designers into a constrained

---

3 The suffix 'friendly' has been attached to so many nouns that it is easy to forget that it came to prominence in recent times as an ergonomic descriptor of software. I have seen various processes, objects and events described in the past few years as 'eater-friendly' (a variety of bean!) 'customer-friendly' (a shop), 'audience-friendly' (a theatre) and 'cyclist-friendly' (a road!), and it is clear that the philosophy of user-centredness has at least been hijacked by marketing professionals.
4 WIMP is an acronym for Windows, Icons, Menus and Pointing devices (or alternatively, Windows, Icons, Mouse and Pull-down menus).

view of what is possible with interfaces, and can be blamed in large part for the stunted growth of innovative interfaces in twenty-first-century designs (the Web, in particular, being remarkably homogeneous in form and style for an apparently revolutionary technology adopted and endorsed by free-thinkers).

Operational definitions more accurately reflect the underlying dynamics of HCI and for this reason are preferred by specialists in user experience design. Shackel (1991) states that the usability of a system can be defined as:

> [its] capability in human functional terms to be used easily and effectively by the specified range of users, given specified training and support, to fulfill a specified range of tasks, within the specified range of environmental scenarios.
>
> (p. 24)

The key aspects are 'easily' (that is, the user – with or without training as specified – must not find it difficult to utilise the functionality of the system to perform a task) and 'effectively' (that is, performance must be of a suitably high level, however defined). Many systems could satisfy one of these criteria and either be easy to operate but not very effective in terms of task performance or be extremely difficult to use but, if ever mastered, could support very effective task performance. Neither of these would be deemed very usable in terms of the above definition.

The International Standards Organisation (ISO) introduced standards for information technology (for example, ISO 9241) that cover the issue of software usability and use a slightly modified definition to Shackel's, equating usability with three specific criteria, effectiveness, efficiency and satisfaction, measured in specified contexts. The emergence of usability at the level of international standards for design is testimony to the growing importance of such ergonomic issues in technology production.

The effectiveness, efficiency and satisfaction approach is attractive. It states that at least three measures are important for any given context of use (where context means the types of user and types of environment). Typically, effectiveness is equated with task completion, efficiency with time, and satisfaction is the user's general liking or disliking of the interface. However, the measures can be broader and more sophisticated than this, as shown in Table 2.1.

Each of these criteria is negotiated in the context most appropriate for the design. Thus, rather than assuming all users must be 100 per cent effective, efficient and satisfied before we can deem a product usable, the operational approach suggests gathering such measures as a basis for comparison with existing tools or as a means of setting reasonable targets for design teams to meet. For example, we might consider the development of a new online textbook for teaching and assess its usability by comparing student performance against an existing paper book. In such a case, relative

Table 2.1 Operational measures of the three usability criteria common to most
definitions

Usability criterion	Operational measure
Effectiveness	Number of tasks completed Proportion of tasks completed Quality of tasks completed
Efficiency	Time to complete tasks Cognitive effort Cost of completing tasks
Satisfaction	Statements of like/dislike Preference Formal questionnaire scores

usability could be derived by examining such data as time taken to locate
target material in the book (measuring effectiveness in terms of success in
location and efficiency as how long the average student took to find it). This
would give us one basis for comparison, but it would not be the only one.
We could examine time taken to read chapters on screen or on paper (an
efficiency score), level of comprehension attained (an effectiveness score),
preference for screen or paper versions (a satisfaction score), etc. Most likely
we would determine that there is no simple usability metric that applies for
all contexts and we would then select appropriate criteria and levels for our
context. This point is central to any understanding of usability since
comparisons across contexts are only possible to the extent that criteria and
user or task factors are similar.

Within the operational definition of usability the issue of context is
obviously critical. It makes no sense to describe a tool or technology as usable
or unusable in itself. Any tool is made for use by certain users, performing
particular tasks in specific environments. Its usability can only be meaningfully evaluated in relation to such contextual variables. There is no
usability metric that can be employed independently of context since it is the
dynamic interplay of users, tasks and settings that determines efficiency,
effectiveness of use and satisfaction, etc. Thus, any statement of a product's
usability must be made in relation to the user, task and environmental
types for which it is designed. One must therefore view with suspicion
any claim by a design company to have developed 'the most usable product'
of its kind if the accompanying contextual details are not provided. A more
correct statement would be, for example, 'the most usable product for
able-bodied data-entry clerks working in a typical office environment', or
'the most usable product for older users without experience of computers,
working at home' (although even these statements leave much information
unspecified). Unfortunately, such lengthy descriptors do not make convenient
advertising slogans and it is likely that the appeal to generic 'ease of use' in
marketing campaigns will continue to flourish.

The contextual aspect of usability also means that it is incorrect to assume that, by virtue of reference to the implicit ease-of-use issue, all usable products must necessarily be simple to operate. Some technologies are only designed for specialist use and they may be applicable only to complex task domains. In such scenarios, usability for specialists will almost certainly not result in a product that is simple to use for anyone else. Indeed, the product might even be totally unusable by such nonspecialists precisely because it has been designed to be usable in other contexts.

This is also the case with learnability. It is wrong to assume that all usable technologies must necessarily be simple to learn to use. Some tasks require such a level of skill and knowledge to perform (open-heart surgery comes to mind!) that it would not necessarily mean an instrument the expert user found usable (in this case the surgeon) must be easy for anyone else to learn to use effectively or satisfactorily. This is a point that is rarely grasped by those concerned simply with putative ease of use and its assumed dependence on the presence of certain features in the human–computer interface such as windows, graphics and menus. It is possibly such misunderstanding of the real meaning of the concept that underlies some of the ergonomics profession's criticisms of the usability engineering movement. But even when we do understand the concept in operational terms, there are other legitimate concerns that any user experience professional might have, and these are worth exploring here.

Extending the concern with usability: affect in interaction

While the usability engineering approach has gained popularity throughout the 1990s, with the emergence of the Web and the wider diffusion of information technology into all our lives it is clear that there is more to the human response to information technology than is captured under the classic understanding of usability.

Recent research shows that despite performance, users sometimes like or prefer systems that do not yield optimal performance for them (Bailey 1993). What determines this effect is not clear, but we do know that the manner in which users respond to interactive devices is not based solely on concerns with efficiency, for example. Tractinsky (1997) showed that some users' responses to interfaces were significantly influenced by how aesthetically pleasing users rated the design. Follow-up work on this in my laboratory has shown that when asked to rank an interface for usability, the correlation with rankings for visual appeal or attractiveness is almost perfect (Dillon 2001). Interestingly, there appears to be no significant correlation between these sets of rankings and actual performance, at least in the initial stages of interaction, supporting the notion that users' first responses are not always a good indication of how well they will utilize a given system or application.

While this remains an area that requires much further research, it is clear that our analyses of human responses to new designs need to consider more than the classic usability engineering responses of effectiveness and efficiency. As HCI professionals seek to have greater impact on the development and implementation of new technologies, it is likely that even operational definitions will need to be extended to accommodate factors in acceptance that are as yet poorly understood by researchers (see Dillon and Morris 1996 for extended discussion of the acceptance literature).

The concept of usability is sufficiently broad for it to cover many aspects of contemporary research in HCI. One distinct strand involves the formalisation of its definition and its evaluation methods. Another is the analysis of what attributes of an artifact affect its usability. It is important to distinguish these, as their merging under one heading is likely to blur important conceptual and methodological distinctions and lead to confusion in discussion of usability and the human impact of technology. These distinctions are explored briefly in the following sections.

Usability evaluation

It is generally accepted now that interactive technologies cannot be designed from first principles so as to guarantee their usability and that systems need to be tested in some way. The term 'usability evaluation' covers a diverse range of activities all concerned with assessing the likely user response to a proposed design. Most, if not all, design processes now incorporate some form of evaluation, though the precise terms used vary from usability test to user experience analysis, or from customer response analysis to user reaction assessment. Usability evaluations, however titled, generally take one of three forms: model-based, expert-based or user-based.

Model-based evaluations

Model-based approaches to usability evaluation are the least common form of evaluation but several methods have been proposed that can accurately predict certain aspects of user performance with an interface, such as time to task completion or difficulty of learning a task sequence. In such cases, the evaluator determines the exact sequence of behaviours a user will exhibit through detailed task analysis, applies an analytical model to this sequence and calculates the index of usability.

The most common model-based approach to estimating usability is the GOMS (Goals, Operators, Mehods and Selection rules) method of Card *et al.* (1983), a cognitive psychology-derived framework that casts user behaviour into a sequence of fundamental units (such as moving a cursor to a given screen location or typing a well-practised key sequence), which are allocated time estimates for completion based on experimental findings of human performance from psychology. In this way, any interface design can be analyzed to give an estimate of an expert user's time to complete a task. The model has shown itself to be robust over repeated applications (see for example, Gray *et al.* 1993), though it is limited to predicting time and, only then, for error-free performance in tasks involving little or no decision making.

Expert-based inspections

Expert-based methods refers to any form of usability evaluation that involves an HCI expert examining the application and estimating its likely usability for a given user population. In such cases, users are not employed and the basis for the evaluation lies in the interpretation and judgment of the evaluator. There is considerable interest in this form of evaluation since it can produce results faster and presumably cheaper than user-based tests.

In HCI, two common expert-based usability evaluation methods are 'heuristic evaluation' (for example, Nielsen 1992), and 'cognitive walk-through' (Lewis and Wharton 1997). Both methods aim to provide evaluators with a structured method for examining and reporting problems with an interface. The heuristic method provides a simple list of design guidelines that the evaluator uses to examine the interface screen by screen and while following a typical path through a given task. The evaluator reports violations of the guidelines as likely user problems. In the cognitive walkthrough method, the evaluator first determines the exact sequence of correct task performance, and then estimates, on a screen-by-screen basis, the likely success or failure of the user in performing such a sequence. In both methods, the expert must make an informed guess of the likely reaction of users and explain why certain interface attributes might cause users difficulties.

These methods differ in their precise focus. Heuristic methods are based on design guidelines and ultimately reflect the expert's judgment of how well the interface conforms to good design practice. The cognitive walkthrough method concentrates more on the difficulties users may experience in learning to operate an application to perform a given task. In practice, usability evaluators tend to adapt and modify such methods to suit their purpose and many experts who perform such evaluations employ a hybrid form of the published methods.

User-based tests

Testing an application with a sample of users performing predetermined tasks is generally considered to yield the most reliable and valid estimate of an application's usability. Performed either in a usability test laboratory or a field site, the aim of such a test is to examine the extent to which the application supports the intended users in their work. Tightly coupled to the operational approach to usability definition, the user-based approach draws heavily on the experimental design tradition of human factors psychology in employing task analysis, pre-determined dependent variables and, usually, quantitative analysis of performance supplemented with qualitative methods.

In a typical user-based evaluation, test subjects are asked to perform a set of tasks with the technology. Depending on the primary focus of the evaluator, the users' success at completing the tasks and their speed of performance may be recorded. After the tasks are completed, users are often asked to provide data on likes and dislikes through a survey or interview, or may be asked to view with the evaluator part of their own performance on video and to describe in more detail their performance and perceptions of the application. In this way, measures of effectiveness, efficiency and satisfaction can be derived, problems can be identified and redesign advice can be determined. In certain situations, concurrent verbal protocols might be solicited to shed light on users' thought processes while interacting with the tool so that issues of comprehension and user cognition can be addressed. In a usability laboratory, the complete interaction is normally video recorded for subsequent analysis of transactions, navigation, problem handling, etc. However, more informal approaches are also possible. Some user-based tests are unstructured, involving the user and the evaluator jointly interacting with the system to gain agreement on what works and what is problematic with the design. Such participative approaches can be very useful for exploring interface options in the early stages of design where formal quantitative assessments might be premature.

In an ideal world, user testing with a large sample of the intended user population would occur routinely; however, because of resource limitations, user-based tests are often constrained. As a result, there is considerable interest among HCI professionals in determining how to gain the most information from the smallest sample of users. While popular myths exist about being able to determine a majority of problems with only two or three users, Lewis (1994) has shown that the sample size requirement is largely dependent on the type of errors one seeks to identify and their relative probability of occurrence. Whereas three users might identify major problems in a new application, substantially more users will be required to tease out the remaining problems in a mature or revised product.

Comparisons of methods

The relative advantages and disadvantages of each broad method are summarised in Table 2.2. As usability evaluators are trying to estimate the extent to which real users can employ an application effectively, efficiently and satisfactorily, properly executed user-based methods are always going to give the truest estimate. However, the usability evaluator does not always have the necessary resources to perform such evaluations and therefore other methods must be used.

HCI professionals have recently reported comparisons of evaluation methods but few firm conclusions can yet be drawn. John and Marks (1997)

Table 2.2 Relative advantages and disadvantages of each usability evaluation
method

Usability method	Advantages	Disadvantages
User-based	Most realistic estimate of usability Can give clear record of important problems	Time-consuming Costly for large sample of users Requires prototype to occur
Expert-based	Cheap Fast	Expert-variability unduly affects outcome May overestimate true number of problems
Model-based	Provides rigorous estimate of usability criterion Can be performed on interface specification	Measures only one component of usability Limited task applicability

compared multiple evaluation methods and concluded that no one method
is best and all evaluation methods are of limited value. Andre *et al.* (1999)
attempted a meta-analysis of 17 comparative studies and remarked that a
robust meta-analysis was impossible due to the failure of many evaluation
comparisons to provide sufficient statistics. Caveats noted, several practical
findings follow:

- It is generally recognised that expert-based evaluations employing the
  heuristic method locate more problems than other methods, including
  user-based tests. This may suggest that heuristic approaches label as
  problems many interface attributes that users do not experience as
  problems or are able to work around.
- The skill-level of the evaluator performing the expert-based method
  is important. Nielsen (1993) reports that novice evaluators identify
  significantly fewer problems than experienced evaluators, and both of
  these groups identify fewer than evaluators who are both expert in
  usability testing and the task domain for which the tool under review is
  being designed.
- Team or multiple expert evaluations produce better results than single
  expert evaluations.

Finally, there are good reasons for thinking that the best approach to
evaluating usability is to combine methods, for example using the expert-
based approach to identify problems and inform the design of a user-based
test scenario, as the overlap between the outputs of these methods is only
partial, and a user-based test normally cannot cover as much of the interface

as an expert-based method. Obviously, where usability evaluation occurs throughout the design process, the deployment of various methods at different stages is both useful and likely to lead to greater usability in the final product.

The laboratory versus the field

Another issue in evaluation concerns the use of laboratories and field sites for testing usability. In the 1980s many companies, in a highly publicised attempt to display a firm commitment to usability, invested heavily in usability laboratories to test their products (Dillon 1988). In effect, these laboratories resulted in the generation of a lot of videotape of users interacting with tools in simulated offices but in the main served only to highlight the shortcomings of evaluation methodology rather than the advance of usability engineering. In fact, a major weakness of the user-centred design philosophy has been the ease with which one can claim adherence to it without a commensurate shift in the design process (Dillon *et al.* 1993).

As laboratory studies have often failed to predict real-world usability, a move towards taking the contextual nature of usability to its limits has emerged and some proponents claim that usability can only be assessed in the field. This perspective is supported by research that shows how some laboratory evaluations have failed to pick up usability problems that have been found in the field (see for example, Bailey *et al.* 1988). However, while the ultimate proof of a design is its reception in the real world as demonstrated over time, field tests, like laboratory trials, are only as good as the evaluator who plans and performs them. Such findings say less about the problems of laboratory evaluations than about the assumptions these evaluators made on tasks performed with the tool, which were subsequently shown to be false in the field. As has been stated, evaluations must occur in context and if the context can be reproduced accurately in a laboratory evaluation then all well and good.

What makes a technology more or less usable?

The process of evaluation is important but in itself tells the design team little about how to improve usability or design it into the system at the earliest stages. Making the leap from evaluation result to design attribute that influences usability requires knowledge of the psychology of HCI, the literature on interface design in general, and specific knowledge of the application domain in which the designed artifact is located.

Relevant psychological characteristics of the interaction process include the information processing characteristics and limitations of the user such as the tendency to interpret information in terms of pre-existing knowledge structures or schemata, the selective nature of attention, the nature of visual, auditory and haptic perception, the limitations of short-term memory and so forth. It is not the intention of the present book to cover material such as this, which is dealt with at a range of levels and breadths in a vast existing literature (see for example, Gardiner and Christie 1987 for a good introduction to HCI-related psychology).

The literature on interface design can be traced back over 20 years and in that time innumerable interface features have been put to empirical evaluation and comparison. In one sense, this work reflects a flawed research paradigm that sought to identify the core set of 'good' interface features that could be used as building blocks for interface design. However, this is an extreme view and few ergonomists or human factors professionals carrying out such comparisons really believe such a 'recipe-book' approach is now feasible, not least because of the situated or contextual nature of HCI. Nevertheless, the literature contains many guidelines based on work which purports to show that, say, menus are better than command languages, large windows are better than small windows or the mouse is better than function keys for input, much of which is extremely useful in designing user interfaces but *only* if appropriately interpreted. The relevant aspects of this literature for the designer of electronic texts will be critically evaluated in the next chapter. Now, it is important to place this discussion of usability within the wider process of systems design.

Towards effective user-centred design processes

Traditional models of the design process make reference to numerous stages such as conceptual design, specification, designing, building and testing, through which a product is seen to advance logically. In software design, structured decomposition was the dominant paradigm as manifest, for example, in the waterfall model – so called since activities are supposed to be completed at one stage before the next stage commences, and there is no need (or scope) to move backwards in the process (see Boehm 1988). In the context of such processes, the traditional role of ergonomics professionals has been to develop suitable test procedures for the evaluation stage as outlined earlier. Given what has been said of usability thus far, it is easy to see the relevance of the ergonomic approach to that stage. However, this has long been a source of complaint by user specialists who have rightly pointed out that by the time they receive a product for evaluation there is little chance that they can persuade a company to make drastic alterations despite sometimes-strong evidence of usability problems. Furthermore, problems identified at this stage could often be traced back to faults that user specialists claimed they would have been able to identify far earlier in the process if only they had been consulted. However, if one cannot accurately prescribe usable interfaces in advance (which we cannot) then a problem exists in terms of the role user specialists can expect to play at the pre-evaluation stages of the design process. Being seen as evaluators of (other) designers' work, and consulted only when others see ergonomics as having something to offer, is guaranteed to limit the profession's input.

In effect, the whole user experience movement has sought to resolve this dilemma by advocating an alternative process for artifact design that is generically termed 'user-centred' (see for example, Norman and Draper 1986; Eason 1988). User-centred design advocates a process that will increase the chances of an acceptable product being produced even if in advance it cannot be completely specified. The general philosophy is one of design–test iteration within the product life cycle, in contrast to more traditional phased design models that emphasise the logical progression of products from specification to design stages, to building and finally testing.

In reality, design is rarely the smooth linear sequence of developmental stages many models suggest (except maybe in the eyes of senior project managers or financiers). Rather it is an iterative sequence of events consisting of many subcycles and complex, often poorly understood and certainly opportunistic, processes involving aspects of all stages. It is precisely this cyclical and opportunistic character that user-centred design methods seek to exploit. The fundamental premise of user-centred approaches (as the name suggests) is the addressing of user-relevant issues at all stages in whatever form is appropriate. Thus, assuming a generic representation of the design process as the eventual progression from specification, conceptual design, physical design, evaluation and release (with apologies to advocates of other 'key' stages), then the user-centred approach is to find a means of addressing relevant user issues at each and every stage.

Proponents of user-centred design suggest that it is insufficient just to wait for a usability evaluation late in the design process to gain feedback on user issues. Instead they argue that a system's eventual success is largely related to the degree to which user issues are addressed early in the design process. It is easy enough to see how standard usability evaluation processes can be applied as soon as some representation of the product is available to be tested. With the advent of cheap and fast prototyping applications it is now possible to mock-up a design for an interface rapidly and then test it and redesign accordingly, thereby effectively removing the problem of only influencing the design process when it is too late to change anything. Little more needs to be said of this. Rapid prototyping is a major step forward and it should be simple enough to see how all that has been said so far in this chapter about usability can be applied to prototypes. However, there is more to user-centred design than just bringing the standard evaluation role forward in the product life cycle and many human factors professionals are seeking to impact the very earliest stages of design that lead to the specification of the product prior to any prototype of it.[5]

---

5 Once could add a host of other reasons why user-centredness should not be merely equated
   with rapid prototyping such as the changing form of user requirements over time, the

One can identify variants of user-centered systems design by the nature of the user's involvement in the process. While we all advocate user involvement, the manner of involvement can vary, from test subject (the classic usability engineering approach), through participant in design decisions, to collaborator in studying and conceiving designs that will work. Despite the generic name, user-centered design seems to have different users at its centre, depending on the philosophy of the major stakeholders involved.

Theoretically and practically, earlier involvement is necessary, since the ability to produce a cheap and quick model or prototype of the system is not itself logically sufficient to produce usable designs. Prototypes might continually be built and tested but not get near acceptable levels of usability if information about the target users and their tasks is weak or incorrect. Thus, according to user-centred principles, ergonomics needs to have impact at the stage prior to any building of the product, that is, the requirements and conceptual design stages. Here, ergonomists seek to influence the very conception of the design and how it might look in accordance with informed views of the target users and their tasks.

One important human factors input for the earliest stage of design is user requirements analysis. This can involve a set of procedures such as stakeholder and scenario identification, and user and task analyses, the basic aim of which is the provision of as much relevant, accurate information as possible about user-related issues that are likely to influence the eventual usability of a system. This information might be general or specific but it should act to constrain the range of design alternatives to consider and identify any crucial shortcomings in the subsequently proposed system. In this sense, the first prototypes are influenced for the better and the number of design–test iterations can be reduced.

Stakeholder identification

Stakeholders are all the key people who have an interest in the system to be designed. Thus stakeholders include not only the target users but also those designing and maintaining the system, those offering user-support such as trainers and system managers, and those whose work might be affected indirectly by the system, for example recipients of output. This is particularly important in organisations where firm boundaries between direct and indirect users of a system and its outputs are blurred. For some applications the target end-users are the only key people involved, for others the stakeholders are many.

Each stakeholder should have the various events that could occur in their

emergence of other requirements only when a system has been used for real, the absence of functionality in prototypes, the validity of testing prototypes and the tendency for designers to preserve rather than reject prototyped interfaces.

use of the system identified. This is termed scenario identification. It naturally emerges for direct end-users during task analysis but is often difficult to predict in advance for all stakeholders in the technology. Gardener and McKenzie (1988) recommend considering scenarios in terms of criticality, difficulty, frequency of occurrence and any other distinctions viewed as relevant to that application. Ideally all scenarios should be explicitly considered in the design specification but this is invariably impossible. However, the process can be used to identify the critical ones to address for each stakeholder.

Scenario identification

A scenario is a story about an interaction a user experiences with an information resource which conveys explicit information about what a user does in a specific context, how s/he experiences the interaction, what the goals are and what actions are taken. Scenarios have plots, sequences, events, etc., and are created by the design team, often with input from the users and other stakeholders, to allow exploration of the requirements for a new resource.

At the earliest stages of design, scenarios can prove useful for establishing what is needed without focusing undue attention on how a design might be accomplished technically. By allowing speculative imaginings of use and users, the process enables all participants to share their ideas, at least in theory, and come to a common understanding of what the goals of the design process should be (Rosson and Carroll 2001). In practice, scenario exploration touches on user and task analysis as well as stakeholder discussions and the term is often used to describe a process more than a stage.

User analysis

User analysis involves identifying the target end-users of the system and describing them in as much detail as possible in terms relevant to system usage. This might consist of any particular skills they have or need to be trained to acquire and the possible constraints these might impose on system effectiveness. It might also include identification of particular constraints imposed by working/usage environments such as noise and light conditions, which, though not strictly user characteristics, might often interact sufficiently with user performance to be considered a relevant factor.[6]

In the early years of human factors work on computer design various attempts were made to classify users into types in the hope that an acceptable

---

6 An example might be a distinction between users in terms of 'office' or 'shopfloor' worker. These descriptors appear to distinguish users on little other than a label but in fact inform us of potential environmental constraints that one should examine closer.

classification system would enable accurate predictions about usability to be made at the outset of design. Unfortunately, many of these failed to demonstrate any validity and currently few typologies based on individual psychological differences (for example personality, cognitive style, etc.) can be shown to have any relevance to the earliest stages of software design. However, user analysis has utility in making explicit any assumptions about end-users and their characteristics and works at the specific context level to ensure an accurate image of the user is shared by the design team (see Dillon and Watson 1996).

Task analysis

Task analysis may be defined as the process of identifying and describing units of work and analysing the human and equipment/environmental resources necessary for successful work performance. It is often used synonymously with task description, job analysis, work study, etc., which is not accurate as each of these terms refers to techniques that share some aspects of task analysis but ultimately include issues that are insufficient for, or distinct from, pure task analysis. In the present context I take the term to imply the analysis, synthesis, interpretation and evaluation of task requirements in the light of knowledge and theory about human characteristics.

Historically, numerous techniques for task analysis have been described (for example, Annett and Duncan 1967; Miller 1967; Moran 1981). Broad distinctions can be drawn between them in terms of level of detail addressed (the 'grain of analysis'), formality of procedure involved and expertise required by the analyst. The basic principle common to all such techniques, however, is the identification of procedures performed by the human during the task and the subsequent decomposition of these into their constituent cognitive, perceptual and physical elements. Psychological knowledge is brought to bear on the analysis and inform on optimal organisation and synthesis of the cognitive and perceptual elements while physiological knowledge is related to the physical elements where relevant. With software, the cognitive ergonomic issues are primary, whereas physiological issues are mainly related to the hardware side of information technology. There is not always such convenient separation. For example, input devices are essentially a hardware issue for which cognitive aspects are sometimes discussed.

In the HCI domain analysis techniques are continually emerging to aid designers of interactive computer systems, although the term 'analysis' in this context is a bit of a stretch and has recently been much criticised. Despite the advantages of task analysis at the specification stage there are several shortcomings in the approach that need to be appreciated. The obvious one relates to the level of analysis to be employed. Most human factors practitioners prefer general descriptive levels of analysis rather than fine-grain numerical ones (such as the Keystroke level of the GOMS approach). This probably relates to human factors' strengths, which lie in qualitative rather than quantitative analysis, and evaluation rather than prediction. This might not be such a problem but for the fact that computer scientists and software engineers strongly distrust human factors' apparently vague suggestions and guidelines. Overcoming this obstacle is not simple and it accounts for many of the difficulties user specialists have had in influencing designs at the earliest stage.

Less obvious a problem (but one central to the design of electronic documents and hypermedia), however, is that of employing task analysis for situations where innovative products are being developed. By definition, such systems are likely to require skills and procedures for which existing tasks are not directly comparable. In a similar way, building less innovative systems solely on the basis of observed current task performance is likely to limit the potential for improvement and innovation offered by designing software to support the task since it locks in old methods to new tools without justification. At a cognitive level, basic representations of the task space need to be maintained to facilitate positive transfer of user knowledge to any new system for similar work. The important point, however, is to identify the crucial elements to maintain and use them to inform the design of any more innovative aspects. Obviously, detailed knowledge of cognition is a prerequisite of such analysis. Even so, such knowledge would hardly ever be enough.

This point further highlights the need for iteration within the design cycle. As stated earlier, design does not proceed through a finite series of discrete stages but rather requires iteration around the basic triumvirate of specification–design–evaluation. For innovative products suitable task analysis can inform early on but should be backed up with simulations and prototypes for evaluation as soon as possible thereafter. Only in this way can designers be sure of achieving their goal. This point is elaborated further in Shackel (1986) and Diaper (1990). It is unlikely that an ergonomically correct interface will be derived from the first analysis of usability requirements. Rather such work will act to constrain the number of design choices before further analysis and evaluation work is carried out which moves the design closer to its final form. The limitations of task analysis points to the need for alternative means of exploring options, such as scenario analysis, mentioned earlier.

From analysis to specification and prototype

The leap from information gathered at the earliest stages of design to system specification and subsequent prototyping cannot always be logically determined. For fine-grain task analyses (such as a keystroke level analysis) the results can logically dictate the choice of design option. However, as most analyses are higher level and some are purely descriptive, such logical derivations concerning usability cannot often be made. However, good task analyses, even in the design of innovative systems, can be used to decide between several interface alternatives. For example, knowing that users will be reading large amounts of text on screen could be related to the human factors literature on these issues, which offers advice on screen size, resolution, manipulation facilities and so forth, thereby constraining the range of suitable alternatives to be considered. From task and scenario analysis at least, we might expect a mapping from the described user cognitions and behaviours to some specific aspects of the interface.

Stakeholder identifications and user analyses provide less specific routes to interface specification. Generally, they inform the process of information gathering at this stage and can be used as supporting evidence to guide decisions taken primarily on the basis of task analysis. Firm boundaries are difficult if not impossible to draw here. The information obtained from any procedure is often similar to that obtained from others, the differences occur in the degree to which the technique concentrates on specific aspects of the users. Ultimately, however, any mapping from analysis to interface specification requires some knowledge of human factors.

This nonprescriptive approach to design can be problematic in examining the precise role ergonomic knowledge and methods may have in the design process. Even with the best information available on users, their tasks and the context in which they will perform these tasks, there are still many potential ways in which a designer could proceed. User-centred approaches suggest that the designer should produce a prototype solution and then test this, using the subsequent results to inform the further design of the product until agreed or target levels of usability are achieved.

However, there are potential shortcomings here that need to be addressed. As has been pointed out earlier, it is unsatisfactory merely to propose that prototypes are tested until they get better because, theoretically, we have no guarantees they will ever be acceptable. Certainly they should, but the process of improvement is contingent on appropriate testing, which has been outlined earlier as a process best left to people who know about running effective usability trials. Evidence from the European software industry (Dillon *et al.* 1993) suggests that all too often, testing is carried out informally by the designers themselves with minimal or no user involvement. In attempting to change design processes in order to maximise ergonomic input at this stage, the profession is in danger of creating a process with little or none.

Furthermore, unless we have some idea of what the designers are doing with the requirements information in coming up with a prototype – in short, understanding design as a social and cognitive process – we are always missing a major variable in the equation (see for example, Curtis 1990). HCI has attempted to provide designers with guidelines and data in order to influence design for the better, but the impact has been minimal. As a field it can rightly be criticised for its failure to address adequately the nature of design at these levels and it is the author's contention that progress towards truly effective user-centred design processes (and they will be processes as we cannot describe one universal process to cover all artifacts or design tasks) will be hampered until we have a better understanding of these issues.

Design and science: the product as conjecture

Given a set of requirements and/or information about typical usage, what does a designer or design team do next? This is a deceptively simple-looking question to which many would offer the answer: 'they produce a prototype' (or a 'design', a 'model', a 'partial solution', etc.). Yet such an answer tells us little or nothing about the activity of design at this level. Investigations of designers designing are relatively sparse in the literature although the decisions taken at this stage must have an impact out of all proportion to their effort in the product life cycle.

Work with engineers and architects has shed some light on the process. Not surprisingly, engineering design has tackled this issue for many years and there are numerous models for effective design in the literature (see for example, Archer's 1965 'linear, critical-path approach' or Asimow's 1962 'design morphology'). These tended to be prescriptive and sought to guide the designer towards a successful solution rather than reflect the cognitive processes involved. More recently, however, the design task (not only or even predominantly software design) has been examined typically as a form of problem solving amenable to analysis by cognitive science.

Part of this literature seems to have uncritically adopted much of the work on cognitive style from mainstream psychology. Cognitive style can loosely be defined as the manner in which people process and respond to information and rests on an assumption that individuals can be distinguished in terms of characteristic processing and response. This has led to the proposal of numerous dimensions of style such as field dependence–field independence (Witkin *et al.* 1977), convergence–divergence (Hudson 1968) and holism–serialism (Pask 1976). Few of these putative styles have been shown to predict design performance reliably (or much else for that matter) but the idea of cognitive style remains seductive and has given birth to the notion that designers might manifest distinctive styles of reasoning that could be identified and used to aid design education and to develop computerised-aided design (CAD) tools (see for example, Tovey 1986; Cross 1985).

Empirical work on design has sidestepped much of this theorising. Current thought frequently emphasises the generation–conjecture–analysis model of design originally proposed by Darke (1979). She interviewed successful architects about their own way of designing and concluded that designers proceed by identifying (often through their own experiences and values as much as by stated requirements) major aims for the design that she terms 'the primary generators'. On the basis of these, the designer proposes a partial solution (the 'conjecture') that is subsequently analysed in the light of requirements before the process is repeated. Interestingly, she reported that designers in her sample (admittedly only seven) rarely waited for the requirements to be worked out in detail before producing a conjectured solution, most claiming that only with such a solution was it possible to identify many important requirements.

Lawson (1979) compared student scientists with student architects in a problem-solving task involving the arrangement of coloured cubes into a form that would satisfy a stated criterion. He reported that distinctive problem-solving strategies could be identified between these groups. Scientists tended to consider all options before attempting logically to determine an acceptable solution (the so-called logico-deductive approach). Architects adopted a more impulsive strategy, proposing numerous solutions before they completely understood the problem and using the solution as a means to find out more about the target criterion.

Such work is often taken to suggest that designers possess a unique form of cognitive strategy that distinguishes them from scientists, but such an argument is less easy to maintain if the context of each group's work is considered more deeply. Designers are judged by their solutions more than their reasoning so it is natural that they seek to demonstrate proposed solutions. Scientists tend to be judged at least in part on their reasoning as much as their answers, and the quality of the answer itself is often judged in terms of the reasoning that led to it. Perhaps scientists deal with more abstract entities and merely examining the outputs of scientific work tells us little enough about their intrinsic worth. With designers dealing with physical objects, a proposed solution can quite quickly communicate its own worth directly to an observer. Thus, the strategies manifest in such studies may reflect task variables and experience more than underlying cognitive biases or information processing styles.

What is most interesting is Darke's use of the term 'conjecture' to describe a designer's initial solutions. This invokes Popper's description of science as the process of conjecture and refutation by which human knowledge accrues. Contrary to the popular image of scientists as rigidly logical, clinically objective, methodical planners (as seems to be suggested by those who (mis)interpret Lawson's oft-cited findings), Popper argues that scientists do, in fact, guess, anticipate and propose tentative solutions that are then subjected to criticism and attempted refutation. According to this philosophy of science, proposed answers and stern tests of their suitability are our only sure way of making progress.

In this way, designers and scientists are less different than many would seek to advocate. What may separate them is less some supposed cognitive characteristic such as style or problem-solving strategy but more the degree to which they seek to rationalise their conjectures in advance and evaluate them once produced. Science has long sought to justify its conjectures in terms of precedent, predicted outcome and theoretical rigour. Similarly, criticism of conjectures acts as a reality check to hold speculative conjectures in perspective. Indeed, from a Popperian perspective, if a theory cannot be put to the test such that it can be proven wrong, the theory cannot even be called 'scientific'.

Design practice thus seems similar to science so construed. By extension, a design solution is really just a hypothesis; a conjecture on the part of the designer as to what will suit the intended users or meet the specified requirements. The usability test provides the practical means of refutation. Evidence may suggest that the designers are less rigorous in their test methods than many human factors professionals (here masquerading as scientists!) would like them to be, but if we accept Popper's model of science, they are both performing very similar activities. Design may be more craft than science but the demarcation between these activities is often subtle and does not deny their similarity. As Popper (1957) says:

> there is no clearly marked division between the pre-scientific and the scientific experimental approaches, even though the more and more conscious application of scientific, that is to say critical methods, is of great importance. Both approaches may be described, fundamentally, as utilising the method of trial and error. We try, that is, we do not merely register an observation but make active attempts to solve some more or less practical and definite problems. And we make progress if, and only if, we are prepared to learn from our mistakes: to recognise our errors and utilise them critically instead of persevering in them dogmatically. Though this analysis may sound trivial, it describes, I believe, the method of all empirical science. This method assumes a more and more scientific character the more freely and consciously we are prepared to risk a trial and the more critically we watch for the mistakes we always make. And this formula covers not only the method of experiment but also the relationship between theory and experiment. All theories are trials; they are tentative hypotheses, tried out to see if they work; and all experimental corroboration is simply the result of tests undertaken in a critical spirit, in an attempt to find out where our theories err.
>
> (*The Poverty of Historicism*, p. 87 in 1986 edition)

Seen in this light, science is very similar to the models of design proposed by Darke (1979), Lawson (1979) and others. All seek solutions to problems, all conjecture and refute (even if not to similar degrees), and none can be sure how they ever come up with an idea for a theory, an experiment or a design. If we can conceptualise the design process at the designer level as similar to scientific practice as outlined by Popper, we spend less time searching for uniquely designer-like modes of thought and address the issue of influencing for the better the conjectures (proposed design solutions) and the systems of refutation (the evaluation methods). In many ways, this is the theme of the present book. The design *is* the theory, and the evidence in its support or refutation must come from use.

Electronic documents as a technology?

At the start of this chapter it was stated that electronic texts were just one more application of computers and a relatively natural one at that. Talk of documents as a technology might be disconcerting at first but this probably results more from an old-fashioned view of what a technology is rather than any flaw in the description. Certainly if we can talk of information technology as more than the technical or applied scientific aspects of information handling, in fact as being the physical instantiation of these principles, then a document is a technology for reading and electronic texts are just another example of reading technology.

Who then are the stakeholders and users of this technology? Almost everyone reads and ideally everyone should be able to do so. Parents, educators, governments and certain employers expend vast amounts of capital and effort in ensuring basic levels of literacy are maintained and maximised. To the extent that almost everyone is a reader, then almost everyone is a user or potential user of this technology. And to the extent that society places great store by literacy, everyone is a stakeholder in information technologies.

The very breadth of the user population can mislead some people into thinking that we need not analyse the user population for this technology much further since the users are 'everyone and anyone'. This is a mistake. The diversity of the user population in total means that they embody almost all sources of individual difference imaginable in the human species: language ability, intelligence, experience, task knowledge, age, education, cognitive style, etc. Add to these the cultural differences that exist across groups and societies (about which, as yet, user experience professionals know very little) and it is not difficult to see that the need for more user studies is immense. As stated in relation to contextual issues, the specific nature of the user population any one document is aimed at needs to be made explicit in any discussion of its usability.

We would expect to find users of documents in work situations, schools and universities, homes, public environments and so forth. The type of document will range from technical to leisure material, books and magazines, articles and essays, notes and memos, bills and love-letters, etc. Users will be reading for pleasure and for need, to learn and to amuse, to find specific information and to browse, to pass exams and to pass the time. Thus, when talking about electronic documents and their usability it is essential that the contextual variables are made explicit and we avoid the trap of endorsing or dismissing the medium on the basis of an excellent or an inappropriate design for one specific context. These issues will become clearer as the book progresses and specific document design issues are examined.

3 So what do we know?

An overview of the empirical literature on reading from screens

Of all the needs that a book has, the chief need is that it be readable.

Anthony Trollope, 1883

Introduction

In simple terms, there exist two schools of thought on the subject of electronic texts. The first holds that paper is far superior and will never be replaced by screens. The argument is frequently supported by reference either to the type of reading scenarios that would currently prove difficult, if not impossible, to support acceptably with electronic text, for example, reading a newspaper on the beach or browsing a magazine in bed, or to the unique tactile qualities of paper. The latter aspect is summed up neatly in Garland's (1982) comment that electronic text may have potential uses, 'but a book is a book is a *book*. A reassuring, feel-the-weight, take-your-own-time kind of thing' (cited in Waller 1987, p. 261).

The second school favours the use of electronic text, citing ease of storage and retrieval, flexibility of structure and the saving of natural resources as major incentives. According to this perspective, electronic text will soon replace paper, and in a short time (usually 10 years hence) we shall all be reading from screens as a matter of habit. In the words of its greatest proponent, Ted Nelson (1987), 'the question is not can we do everything on screens, but when will we, how will we and how can we make it great? This is an article of faith – its simple obviousness defies argument. If you don't get it there is no persuading you; if you do get it you don't need to be persuaded'.[1]

Such extremist positions might be entertaining though it is now clear to many researchers that neither is particularly satisfactory. Reading from screens *is* different from paper and there are many scenarios, such as those

---

1 As Nelson's book was distributed as a hypertext document there are no page numbers. However, this quote can be located in Chapter 1, An Obvious Vision, under the heading Tomorrow's World of Text on Screen. Such lengthy reference to a specific section highlights a potential problem with hypertext that must be addressed.

cited, that current technology does not support well, if at all. However, technology is evolving, and electronic text of the future is unlikely to be handicapped by limitations in screen image and portability that currently seem such major obstacles. As Licklider pointed out when considering the application of computers in libraries as early as 1965, 'our thinking and our planning need not be, and indeed should not be, limited by literal interpretation of the existing technology' (p. 19).

Even so, paper is an information carrier *par excellence* and possesses an intimacy of interaction that cannot be obtained in a medium that by definition imposes a microchip interface between the reader and the text. Furthermore, the millions of books that exist now will not all find their way into electronic form, thus ensuring the existence of paper documentation for many years yet (but see Baker 2001 for a withering critique of the archival practices of libraries, which, he claims, are destroying vast amounts of paper without concern for the cultural records that are being lost).

The aim of the present chapter is not to resolve the issue of whether one or other medium will dominate (my own view is that they will coexist for a long time yet) but to examine critically the reported differences between them in terms of use so as to support reasoned analysis of the paper versus electronic text debate from the perspective of the reader. The intention is not to support prejudices or to allow people to 'get' the message (in Nelson-speak) but to let the evidence do the talking. In so doing it indicates the approach taken to date by most investigators of electronic text and thus where such work seeks to lead document designers.

The outline of the review

At the outset it must be stated that drawing any firm and generalisable conclusions from the literature is difficult but not impossible. The problems stem from several sources: methodological differences, research questions, and quality of reporting. Other reviewers have noted these issues since work of this kind was first synthesised. For example, Helander *et al.* (1984) evaluated 82 studies concerning human factors research on computer screens and concluded:

> Lack of scientific rigour has reduced the value of many of these studies. Especially frequent were flaws in experimental design and subject selection, both of which threaten the validity of results. In addition, the choice of experimental settings and dependent and independent variables often made it difficult to generalize the results beyond the conditions of the particular study.
>
> (p. 55)

In 1994 I drew attention to the limited experimental tasks employed by many investigators that had produced a history of flawed or nongeneralisable research. Since then the situation has become further complicated as many researchers have jumped into the fray with studies of reading on the Web that ignore the emerging theory of screen reading that has evolved since the late 1980s and early 1990s. Yet despite these problems, I reiterate my 1994 edict: a review of experimental evidence is always the most appropriate place to start.

A detailed literature already exists on typographical issues related to text presentation on paper (see particularly the work of Tinker cited later) and issues such as line spacing and formatting are well researched. This work will not be reviewed here as much of it remains unreplicated on screen, and evidence to date suggests that, even when such factors are held constant, reading differences between the two presentation media remain (see for example Creed *et al.* 1987). Furthermore, this is not a review of every single variable that has been tested. Rather, it is an attempt to understand differences between media as experienced by users. The organisational structure of the review is represented in Table 3.1.

In the first instance this review examines the range of claimed differences between the media and draws a distinction here between outcome and process differences. Following this, a brief overview of the type of research that has been carried out is presented. Experimental comparisons of reading from paper and screen are then reviewed according to the physical, perceptual, cognitive or social level of analysis adopted by various researchers. A final section highlights the shortcomings of much of this work and indicates the way forward for research in this domain.

Table 3.1 Overview of the literature on reading from screens

Claimed differences between paper and screen reading

Outcomes	Processes
Speed	Eye movements
Accuracy	Navigation
Comprehension	Manipulation
Fatigue	
Preference	

Explanations of the claimed differences at four levels

Physical	Perceptual	Cognitive	Social
Orientation	Image quality	Short-term	Cultural forces
Aspect ratio	Flicker	memory	Genres
Handling and	Angle of view	Visual memory	
manipulation	Polarity	Schema for text	
Display size	Display	Searching	
	characteristics	Individual	
	Anti-aliasing	differences	

Observed differences: outcome and process measures

Analysing a complex human activity such as reading is not simple and a distinction has been drawn between assessing reading behaviour in terms of outcome and process measures (Schumacher and Waller 1985). Outcome measures concentrate on the results of reading and consider such variables as: amount of information retrieved, accuracy of recall, time taken to read the text, spelling and syntactical errors identified and so forth. Process measures are more concerned with what is happening when a reader uses a text and include such variables as where the reader looks, how she manipulates it and how knowledge of contents is formed.

In the domain of electronic text (and most other interactive technologies) outcome measures take on particular relevance as advocates continually proclaim increased efficiency and improved performance (that is, reading outcomes) with computer-presented material (see Landauer 1995 for an extended critique of such claims). It is not surprising, therefore, to find that most work comparing the two media has concentrated heavily on such differences. However, with the emergence first of hypertext and subsequently the Web, navigation has become a major issue and process measures have gained increased recognition of importance.

This cannot be an exhaustive review as the field is wide and new data are being reported with almost every issue of a relevant journal or proceedings of relevant conferences. The Web is also a breeding ground for unverified and nonrefereed work, and unless I am confident of the quality, I have avoided mention of most of that type. As with the previous edition, this review is intended more as a perspective on this literature and a way of categorising the emerging findings.

Outcome measures

For purposes of organisation, I have categorised the range of outcome measures used by researchers into five types:

- Speed
- Accuracy
- Comprehension
- Fatigue
- Preference

There is nothing sacred about this categorisation. Rather, it represents a convenient and high-level clustering of diverse approaches and multiple authors. As is shown, several studies use more than one of these outcome types in a single experiment. Furthermore, it is possible to operationalise each of these outcomes in more than one way.

Speed

By far the most common experimental finding over the past 20 years is that silent reading from screen is significantly slower than reading from paper (Kak 1981; Muter *et al.* 1982; Wright and Lickorish 1983; Gould and Grischkowsky 1984; Smith and Savory 1989; Leventhal *et al.* 1993; Mayes *et al.* 2001). Figures vary according to means of calculation and experimental design but the weight of evidence suggests a performance deficit of between 20 per cent and 30 per cent when reading from screen. However, this difference has been questioned. Gould *et al.* (1987b) and Muter and Mauretto (1991) have produced findings showing statistically equivalent speed scores for readers from paper and screen, under certain conditions.

As I noted in Dillon (1992), it is not clear whether the same mechanisms have been responsible for the slower speeds observed in these experiments, given the great disparity in procedures. Across studies we find that readers were presented with different types of screen, often varying in resolution, colour of text and background, font and line length. Added to this, the texts they were asked to read might vary from a full screen to multiple pages. In some reports, it is not even clear what the exact presentation details were, as the authors often fail to include them.

The diversity of this literature can be seen in the following examples. Mayes *et al.* (2001) had 40 people read a 19-paragraph popular science article either in the published paper form or in a high-quality scanned form on screen. All participants were given a maximum of 25 minutes to read the article and then complete comprehension and cognitive load questionnaires. Readers using the electronic version took on average more than 4 minutes more than readers of the paper version to read the text. Leventhal *et al.* (1993) examined readers locating information in an encyclopedia presented in hypertext (run on an Apple Macintosh but with no details of screen provided) or paper form. They report a 'marginally significant' effect for speed favouring paper (the figures published indicate that paper readers were on average about 15 per cent faster than screen readers) but the authors stress that the nature of the information location task was crucial.

So, the evidence surrounding the argument for a speed deficit in reading from screen is not simply concluded. A number of intervening variables, such as the size, type and quality of the screen, may have contaminated the results and certainly complicate their interpretation. On balance, it would appear that typical screen reading tends to slow users down. Interestingly, Martin and Platt (2001) show that their user population is so convinced that they are faster with paper that they avoid screen reading where possible. While better screens might reduce these differences, the typical screens that many people now use, and the increasing use of small screens on mobile computing devices, would lend support to that argument that this effect will be with us for several years yet.

Accuracy

Accuracy of reading can refer to any number of everyday outcomes such as locating information on a page, recalling the content of certain sections, correcting spelling mistakes and so forth. In experimental investigations of reading from screens the term 'accuracy' has several meanings too, though it most commonly refers to an individual's ability to identify errors in a proofreading exercise or locate target words in a display. While studies have been carried out that failed to report accuracy differences between screens and paper (for example, Wright and Lickorish 1983; Gould and Grischkowsky 1984), well-controlled experiments by Creed *et al.* (1987) and Wilkinson and Robinshaw (1987) report significantly poorer accuracy for such proofreading tasks on screens.

Since evidence for the effects of presentation media on such accuracy measures often emerges from the same investigations that looked at the speed question, the criticisms of procedure and methodology outlined above apply equally here. The measures of accuracy employed also vary. Gould and Grischkowsky (1984), for example, required subjects to identify misspellings of four types: letter omissions, substitutions, transpositions and additions, randomly inserted at a rate of one per 150 words. Wilkinson and Robinshaw (1987) argue that such a task hardly equates to true proofreading but is merely identification of spelling mistakes. In their study they tried to avoid spelling or contextual mistakes and used errors of five types: missing or additional spaces, missing or additional letters, double or triple reversions, misfits or inappropriate characters, and missing or inappropriate capitals. It is not always clear why some of these error types are not spelling or contextual mistakes but Wilkinson and Robinshaw suggest their approach is more relevant to the task demands of proofreading than Gould and Grischkowsky's.

However, Creed *et al.* (1987) distinguished between visually similar errors (for example, 'e' replaced by 'c'), visually dissimilar errors (for example, 'e' replaced by 'w') and syntactic errors (for example, 'gave' replaced by 'given'). They argue that visually similar and dissimilar errors require visual discrimination for identification while syntactic errors rely on knowledge of the grammatical correctness of the passage for detection and are therefore more cognitively demanding. This error classification was developed in response to what they saw as the shortcomings of the more typical accuracy measures that provide only gross information concerning the factors affecting accurate performance. Their findings indicate that visually dissimilar errors are significantly easier to locate than either visually similar or syntactic errors.

In a widely cited study, Egan *et al.* (1989) compared students' performance on a set of tasks involving a statistics text presented on paper or screen. Students used either the standard textbook or a hypertext version to search for specific information in the text and write essays with the text open.

Incidental learning and subjective ratings were also assessed. The search tasks provide an alternative to, and more realistic measure of, reading accuracy than identifying spelling errors.

The authors report that subjects using the hypertext performed significantly more accurately than those using the paper text. However, a closer look at the experiment is revealing. With respect to the search tasks, the questions posed were varied so that their wording mentioned terms contained in the body of the text, in the headings, in both of these or neither. Not surprisingly, the largest advantage to electronic text was observed where the target information was only mentioned in the body of text (that is, there were no headings referring to it). Here the search facility of the computer outperformed humans, as expected. When the task was less biased against the paper condition, for example searching for information to which there are headings, no significant difference was observed. Interestingly, the poorest performance of all was for SuperBook users searching for information when the question did not contain specific references to words used anywhere in the text. In the absence of suitable search parameters or look-up terms, the hypertext suddenly seemed less effective.

McKnight *et al.* (1990) compared reading in two versions of hypertext, a word processor file and a paper copy of a document on winemaking. The measure of accuracy taken was the number of answers correctly made to a set of questions seeking information to be found in the document. Interestingly, they report no significant difference between paper and word processor file, but readers in both hypertext conditions were significantly less accurate than readers of the paper document. In a subsequent study, McKnight *et al.* (1992) report that readers of a hypertext document complained more than readers of paper about a lack of confidence in finding all relevant material.

Leventhal *et al.* (1993) performed a detailed study of question-answering with an encyclopedia presented as hypertext or paper, and carefully manipulated the type of questions asked in order to ensure a reliable test of distinct information location strategies. Their questions examined factual information located in titles, in the body of text sections or in maps, as well as questions requiring inference from material in the body of the text or in the maps. Interestingly, they report an effect for question type that interacted with presentation medium but no significant effect for medium alone. Readers of the paper version had most difficulties locating information in the body of the text that had no key to location in the question (as in the Egan *et al.* study) but performed best on questions requiring inference and location in maps. All subjects also displayed an improvement with practice that was independent of presentation medium.

Regardless of the interpretation that is put on the results of any one of these studies, the fact remains that investigations of reading accuracy from screen and paper take a variety of measures as indices of performance. Therefore, two studies, both purporting to investigate reading accuracy, may not necessarily measure the same events. The issue of accuracy is further complicated by the presence or absence of certain enabling facilities (for example, search routines) and the potential to alter information structures with hypertext applications. Obviously, for some tasks, the computer is the only viable presentation medium, but the more the task involves serial reading of text and less manipulation and searching, paper retains advantages related to visual discrimination. Before claims are made about the relative accuracy of either medium, clear task descriptions need to be provided.

Comprehension

Perhaps more important than the questions of speed and accuracy of reading is the effect of presentation medium on comprehension. Should any causal relationship ever be identified between reading digital texts and reduced comprehension, the impact of this technology would be severely limited. The issue of comprehension has not been as fully researched as one might expect, perhaps in no small way due to the difficulty of devising a suitable means of quantification; that is, how does one measure a reader's comprehension?

Post-task questions about content of the reading material are perhaps the simplest method of assessment, although care must be taken to ensure that the questions do not simply demand recall skills. Early studies (Kak 1981; Muter *et al.* 1982; Cushman 1986) reported no significant comprehension effect for presentation medium. Belmore (1985) asked subjects to read short passages from screen and paper and measured reading time and comprehension. An initial examination of the results appeared to show a considerable disadvantage, in terms of both comprehension and speed, for screen-presented text. However, further analysis showed that the effect was only found when subjects experienced the screen condition first. Belmore suggested that the performance decrement was due to the subjects' lack of familiarity with computers and reading from screens – a factor commonly found, though rarely controlled adequately, in this type of study.

Very few of the studies reported here attempted to use a sample of regular or experienced computer users, but some such studies do exist. O'Hara and Sellen (1997) tested only experienced users and reported that paper offered several important advantages to users for activities that support comprehension, which these authors 'tested' by having users produce (unscored) summaries.

With the widespread advent of hypermedia, researchers have examined the comprehension issues more systematically. The Egan *et al.* study (1989) required subjects to write essay-type answers to open book questions using paper or hypertext versions of a statistics book. Experts rated the essays and it was observed that users of the hypertext version scored significantly higher marks than users of the paper book. Thus, the authors conclude, the potential of restructuring the text with current technology can significantly improve comprehension for certain tasks.

A study covering this issue by Muter and Mauretto (1991) asked readers to answer questions about a short story read either on paper or on screen immediately after finishing the reading task. They reported no significant comprehension difference between readers using either medium. Similarly, McKnight *et al.* (1992) compared comprehension as assessed by essays written on the basis of notes made during exposure to material presented via hypertext or book. Readers were students taking a postgraduate course of study in this area and a domain expert marked the essays. Again, they reported no significant comprehension difference between the media. However, Leventhal *et al.* (1993) tested incidental learning after their study (described above) by asking subjects to identify titles of sections they had read. They found significantly more incidental learning in the hypertext condition than paper.

Becker and Dwyer (1994) tested undergraduate business students in a beginning course on auditing and computer viruses. The authors developed two treatments, a paper packet and a hypertext program. A pretest ensured a uniform level of background knowledge among students. These authors also found no significant difference between the post-test scores of the hypertext group and the paper group.

Since hypermedia provides a powerful means of manipulating large amounts of data, presumably tasks that require such actions are likely to be better supported in the electronic domain than on paper. Tackling precisely this issue, Lehto *et al.* (1995) compared learner performance on a 'reference' task and a more traditional 'learning task'. After training, 15 graduate students performed two tasks: (1) a 'reading-to-learn' task (which required students to browse and comprehend) and (2) a 'reading-to-do' task (which required students to find and record which annotations contained information on a topic). Two reading-to-learn questions were to be answered using the hypermedia treatment and two using the paper treatment. Time to form the topic, time to answer (summarize), and percentage of relevant references cited by participants were recorded. After completing the reading-to-learn task, the participants were asked to complete 10 reading-to-do questions that were randomly divided between the hypermedia and paper treatments. Time to find relevant references and percentage of relevant references cited by participants were then recorded.

The authors reported that in the reading-to-learn tasks, paper users provided significantly more correct references than did hypermedia users (though this result is reported as significant only at the $p < 0.10$ level). Hypermedia users also took slightly more time to form a topic but slightly less total time to answer questions, though neither of the last two findings were statistically significant. The authors conservatively note that 'taken together, these results clearly do not show an advantage of hypermedia over the book for the reading-to-learn task' (p. 304), though one might add that the significant effect for correct references rather unambiguously suggests a distinct advantage to paper.

Interestingly, the results of the 'reading-to-do' test showed a statistically significant (this time at the standard $p < 0.05$ level) advantage for hypermedia on measures of time to complete the task and percentage of correct references cited. The authors explain this finding in terms of the more flexible search strategies hypermedia make possible though it is not clear from their data that strategy differences explain the findings as elegantly as the speed and power advantages of electronic searching, since the 'to-do' tasks seem to have been little more than word searches. Regardless of the reason, the task dependency of these results suggest one clear direction for further research, location of target information in large documents, as opposed to broad comprehension measured after exposure, where hypertext's increased functionality may offer advantages.

In a major review of the experimental findings on learning outcomes with hypermedia documents, Dillon and Gabbard (1998) concluded that tasks (or subtasks) that involve substantial amounts of large document manipulation, searching through large texts for specific details, and comparison of visual details among objects, are potentially better supported by hypermedia. So it would seem that reading from screens does not negatively affect comprehension rates though it may affect the speed with which readers can attain a given level of comprehension.

Fatigue

The proliferation of information technology has traditionally brought with it fears of harmful or negative side-effects for users who spend a lot of time in front of a screen (see for example, Pearce 1984). In the area of screen reading this has manifested itself in speculation of increased visual fatigue and/or eyestrain when reading from screens as opposed to paper, a claim that was made from the earliest days of electronic text.

In Muter *et al.*'s (1982) study users were requested to complete a rating scale on a number of measures of discomfort including fatigue and eyestrain both before and after exposure to the task. There were no significant differences reported on any of these scales either as a result of condition or time. Similarly, Gould and Grischkowsky (1984) obtained responses to a 16-item 'Feelings Questionnaire' after each of six 45-minute work periods. This questionnaire required subjects to rate their fatigue, levels of tension, mental stress and so forth. Furthermore, various visual measurements, such as flicker and contrast sensitivity, visual acuity and phoria, were taken at the beginning of the day and after each work period. Neither questionnaire responses nor visual measures showed a significant effect for presentation medium. These results led the authors to conclude that good-quality screens in themselves do not produce fatiguing effects, citing Starr (1984) and Sauter *et al.*(1983) as supporting evidence.

In a more specific investigation of fatigue, Cushman (1986) investigated reading from microfiche as well as paper and screens with positive and negative image. He distinguished between visual and general fatigue, assessing the former with the Visual Fatigue Graphic Rating Scale (VFGRS), which subjects used to rate their ocular discomfort, and the latter with the Feeling-Tone Checklist (FTC, Pearson and Byars 1956). With respect to the screen conditions, the VFGRS was administered before the session and after 15, 30, 45 and 60 minutes as well as at the end of the trial at 80 minutes. The FTC was completed before and after the session. The results indicated that reading from positive presentation screens (dark characters on light background) was more fatiguing than paper and leads to greater ocular discomfort than reading from negative presentation screens.

Cushman explained the apparent conflict of these results with the established literature in terms of the refresh rate of the screens employed (60 Hz), which would be considered low today and may not have been enough to completely eliminate flicker in the case of positive presentation, a suspected cause of visual fatigue. Wilkinson and Robinshaw (1987) also reported significantly higher fatigue for screen reading, and while their equipment may also have influenced the finding, they dismiss this as a reasonable explanation on the grounds that no subject reported lack of clarity or flicker and their monitor was typical of the type of screen that users find themselves reading from at that time. They suggest that Gould and Grischkowsky's (1984) equipment was 'too good to show any disadvantage' and that their method of measuring fatigue was artificial. By gathering information after a task and across a working day, Gould and Grischkowsky missed the effects of fatigue within a task session and allowed time of day effects to contaminate the results. Wilkinson and Robinshaw liken the proofreading task used in these studies to vigilance performance and argued that fatigue is more likely to occur within the single work period where there are no rest pauses allowing recovery. Their results showed a performance decrement across the 50-minute task employed, leading them to conclude that reading from typical screens, at least for periods longer than 10-minutes, is likely to lead to greater fatigue.

The question of visual fatigue's widespread existence among users has not been the object of much further experimental work, which is somewhat bewildering given the explosion in screen reading that has occurred in the past decade. However, some research does exist. Sheih (2000) took measures of subjective visual fatigue from readers of liquid-crystal displays (LCDs) while examining viewing distances, polarity and reflection. The users in this study reported higher levels of visual fatigue in situations where there were higher screen reflections. Furthermore, there seemed to be an interesting correlation with viewing distance in that users reported less fatigue as viewing distance increased (which suggests again that multiple variables interact in determining the user response to digital documents).

It is not clear how comparable conclusions drawn from measures of fatigue, such as subjective ratings of ocular discomfort, are with inferences drawn from performance rates. It would seem safe to conclude that users do not find reading from screens intrinsically fatiguing but that performance levels may be more difficult to sustain over time when reading from average quality screens. As screen standards increase over time this problem should be minimised. However, what may take much longer to overcome is the commonly held belief that screen reading is more fatiguing, regardless of any test results from a user study.

Preference

Part of the folklore of human factors research is that naive users tend to dislike using computers and much research aims at encouraging user acceptance of systems through more usable interface design. Given that much of the evidence cited here is based on studies of relatively novice users it is possible that the results are contaminated by subjects' negative predispositions towards reading from screen. Certainly, the quality of 20-year-old screens may account for the results from early studies that reported significant preference for paper (Cakir *et al.* 1980; Muter *et al.* 1982; Starr 1984), but what of more recent work?

Egan *et al.* (1989) found a preference for hypertext over paper among subjects in their study of a statistics text where the electronic copy was displayed on a very high quality screen. Muter and Mauretto (1991) revealed that approximately 50 per cent of subjects in their comparative studies of reading from paper and current screens expressed a preference for screen. A similar figure was reported by Spinks (1991), while Leventhal *et al.* (1993) reported that subjective ratings of task enjoyment were significantly higher for hypertext readers in their comparative study.

Such findings lend some support to the argument that preferences are shifting as screen technology improves or as readers become more familiar with the technology. A recent Pew study (Jones 2002) noted that 85 per cent of university students owned a computer and 73 per cent reported using the internet more than the library for research. So the shift to digital dominance has occurred, but this does not tell the full story. While the OpenEbook forum (the trade organisation for the e-book industry) reports that experience with e-books makes people more willing to use more electronic documents (see http://www.openebook.org), Rho and Gedeon (2000) surveyed users of web-available academic journals and found that only 15 per cent of users read the full text on line, the majority preferring to print the text out for reading, if interested. Martin and Platt (2001) interviewed medical student users and their results indicate a very strong preference still for paper.

What seems to have been overlooked as far as formal investigation is concerned is the natural flexibility of books and paper over screens; for example, paper documents are portable, cheap, apparently 'natural' in our culture, personal and easy to use. The extent to which such common-sense variables influence user performance and preferences is not yet well understood but Martin and Platt (2001) show that such variables really matter to users.

Summary

Empirical investigations of the area have suggested five possible outcome differences between reading from screens and paper. As a result of the variety of methodologies, procedures and stimulus materials employed in these studies, definitive conclusions cannot be drawn. It seems certain that reading speeds are reduced on typical screens and accuracy may be lessened for visually and cognitively demanding tasks. Fears of increased visual fatigue and reduced levels of comprehension as a result of reading from screens have gained little empirical support though the validity of separating accuracy and comprehension into two discrete outcomes is debatable. With respect to reader preference, top quality hardcopy seems to be preferred to screen displays, which is not altogether surprising but shifts may be occurring as technology improves and readers become more experienced with the new medium. As with all interactive experiences, the task being performed is a crucial determinant of usability.

Process measures

Without doubt, the main obstacle to obtaining accurate process data is devising a suitable, nonintrusive observation method. While techniques for measuring eye movements during reading now exist, it is not at all clear from eye-movement records what the reader was thinking or trying to do at any time.

Verbal protocols of people interacting with texts do not require elaborate equipment and can be elicited wherever a subject normally reads. Thus the technique is cheap, relatively naturalistic and physically nonintrusive. However, verbal protocol analysis has been criticised for interfering with the normal cognitive processing involved in task performance and requiring the presence of an experimenter to sustain and record the verbal protocol (Nisbett and Wilson 1977).

Such techniques have rarely been employed with the intention of assessing the process differences between reading from paper and from screen. Where paper and hypertext are compared directly, although process measures may be taken with the computer and/or video cameras, the final comparison often rests on outcome measures (for example, McKnight *et al.* 1990).

Despite this, it is widely accepted that the reading process with screens is different from that with paper regardless of any outcome differences. The following sections outline three of the most commonly cited process differences between the media. In contrast to the outcome differences it will be noted that, for the reasons outlined above, these differences are less clearly demonstrated empirically.

Eye movements

Mills and Weldon (1986) argue that measures of eye movements reflect difficulty, discriminability and comprehensibility of text and can therefore be used as a method of assessing the cognitive effort involved in reading text from paper or screen. Indeed, Tinker (1958) reports on how certain text characteristics affect eye movements and Kolers *et al.* (1981) employed measures of eye movement to investigate the effect of text density on ocular work and reading efficiency. Obviously, if reading from screen is different than paper then noticeable effects in eye-movement patterns might be found, indicating possible causes and means of improvement.

Eye movements during reading are characterised by a series of jumps and fixations. The latter are of duration approximately 250 milliseconds and it is during these that word perception occurs. The 'visual reading field' is the term used to describe that portion of foveal and parafoveal vision from which visual information can be extracted during a fixation and in the context of reading this can be expressed in terms of the number of characters available during a fixation. The visual reading field is subject to interference from text on adjacent lines, the effect of which seems to be a reduction in the number of characters available in any given fixation and hence a reduction in reading speed.

Gould *et al.* (1987a) report an investigation of eye-movement patterns when reading from either medium. Using a photoelectric eye-movement monitoring system, subjects were required to read two 10-page articles, one on paper, the other on screen. Eye movements typically consisted of a series of fixations on a line, with refixations and skipped lines being rare. Movement patterns were classified into four types: fixations, undershoots, regressions and refixations. Analysis revealed that when reading from screen, subjects made significantly more (15 per cent) forward fixations per line. However, this 15% difference translated into only one fixation per line. Generally, eye-movement patterns were similar and no difference in duration was observed. Gould *et al.* explained the 15 per cent fixation difference in terms of image quality variables. Interestingly, they report that there was no evidence that subjects lost their place, 'turned-off' or refixated more when reading from screen.

Since the first edition of this book the largest single study of eye movements during screen reading has been conducted, the Stanford–Poynter Project (http://www.poynterextra.org/et/i.htm). This multiyear project has observed over 60 users, viewing hundreds of news sites on the Web. They have collected over 600 000 eye fixations and sought to identify where users are looking on screen. While there is no media comparison here to allow us to identify sources of difference, the results offer fascinating glimpses of user behaviour online, not least the tendency of users to fixate first on text rather than graphics (in contrast to user behaviour in the paper world).

It is too soon to know what the final analyses of this massive data set will provide but it seems that gross differences in eye movements do not occur between screen and paper reading. However, given the known effect of typographic cueing on eye movements with paper and the oft-stated nontransferability of paper design guidelines to screens, it is possible that hypertext formats might influence the reading process at this level in a manner worth investigation.

Manipulation

Perhaps the most obvious difference between reading from paper and from screens is the ease with which paper can be manipulated and the corresponding difficulty of so doing with electronic text. Yet manipulation is an intrinsic part of the reading process for most tasks. Manipulating paper is achieved by manual dexterity, using fingers to turn pages, keeping one finger in a section as a location aid, or flicking through tens of pages while browsing the contents of a document, activities difficult or impossible to support electronically (Kerr 1986).

Such skills are acquired early in a reader's life and the standard physical format of most documents means these skills are transferable between all document types. With electronic text this does not hold. Lack of standards means that there is a bewildering range of interfaces to computer systems and mastery of manipulation in one application is no guarantee of an ability to use another. Progressing through the electronic document might involve using a mouse and scroll bar in one application and function keys or a stylus in another; one might require menu selection and 'page' numbers while another supports touch-sensitive 'buttons'. With hypertext, manipulation of large electronic texts can be rapid and simple while other systems might take several seconds to refresh the screen after the execution of a 'next page' command.

Such differences will almost certainly affect the reading process. As early as 1988, Richardson *et al.* reported that users find text manipulation on screen awkward compared to paper, stating that the replacement of direct manual interaction with an input device deprived users of much feedback and control. Current hypertext applications, however, support rapid movement between various sections of text, which suggests that innovative manipulations might emerge that, once familiar to users, convey advantages to the reader of electronic texts.

Experimental studies of manipulation differences have concentrated on the effects of scrolling or paging through electronic documents. Popular myths surrounding scrolling characterise much of the discourse on web design where Jakob Nielsen argued (but subsequently revised his opinion) that users did not scroll and missed much information at the end of a web page. Sadly, such proclamations tend to be picked up by many without any attempt to reconcile them with empirical data. Several studies point out disadvantages of scrolling, such as loss of location cues (de Brujin *et al.* 1992;

Piolat *et al.* 1997), yet despite this, users will scroll in web environments (Byrne *et al.* 1999). Dyson and Kipping (1998) suggested that line length might be an important intervening variable since users can and do continue reading even while the text scrolls if the line is short.

O'Hara and Sellen (1997) showed that one of the major advantages of paper documents is their ability to be physically manipulated into different layouts that can aid spatial memory. They suggested that readers of paper exploit the flexibility of paper to create independent reading and writing areas while studying material and that the tactile manipulable properties of paper enable greater reader control of the information space.

Navigation

When reading a lengthy document the readers will need to find their way through the information in a manner that has been likened to navigating a physical environment (Dillon *et al.* 1993). There is a striking consensus among many researchers in the field that this process is the single greatest difficulty for readers of electronic text. This is particularly (but not uniquely) the case with hypertext where frequent reference is made to 'getting lost in hyperspace' (for example, Conklin 1987; McAleese 1989), which is described, in the oft-quoted line of Elm and Woods (1985), as 'the user not having a clear conception of the relationships within the system or knowing his present location in the system relative to the display structure and finding it difficult to decide where to look next within the system' (p. 927).

With paper documents there tend to be at least some standards in terms of organisation. With books, for example, contents pages are usually at the front, indices at the back, and both offer some information on where items are located in the body of the text. Concepts of relative position in the text such as 'before' and 'after' have tangible physical correlates. No such correlation holds with hypertext and such concepts are greatly diminished in standard electronic text.

There is direct empirical evidence in the literature to support the view that navigation can be a problem. Edwards and Hardman (1989), for example, describe a study that required subjects to search through a specially designed hypertext. In total, half the subjects reported feeling lost at some stage (this proportion is inferred from the data reported). Such feelings were mainly due to 'not knowing where to go next' or 'not knowing where they were in relation to the overall structure of the document' rather than 'knowing where to go but not knowing how to get there' (descriptors provided by the authors). Unfortunately, without direct comparison of ratings from subjects reading a paper equivalent we cannot be sure such proportions are solely due to using hypertext.

McKnight *et al.* (1990) compared navigation for paper, word processor and two hypertext documents by examining the number of times readers went to index and contents pages/sections, inferring that time spent there gave an indication of navigation problems. They reported significant differences between paper and both hypertext conditions (the latter proving worse), with word processor users spending about twice as long as paper readers in these sections.

Over the past 10 years navigation has been examined repeatedly in web and related hypermedia designs (Dillon and Schaap 1996; Macdonald and Stevenson 1996; Piolat *et al.* 1997). The general findings suggest that users are more likely to experience navigation difficulties in hypermedia environments than they are in paper and that considerable design effort is required to ensure that the navigational overhead does not limit the cognitive resources available for content comprehension. In a longitudinal study of user adaptation to different versions of online newspapers, Vaughan and Dillon (2000) showed that navigation patterns shift over time, but that poorly designed interfaces, where navigation labels are ambiguous or only vaguely suggest to what material they will lead users, result in slower learning of the information structure underlying the document.

Summary

The reading process is affected by the medium of presentation, though it is extremely difficult to quantify and demonstrate such differences empirically. The major differences appear to occur in manipulation, which seems to be more awkward with electronic texts, and navigation, which seems to be more difficult with electronic documents and particularly hypertexts. Eye-movement patterns do not seem to be significantly altered by presentation medium. Further process issues may emerge as our knowledge and conceptualisation of the reading process improves.

Explaining the differences: a classification of issues

While the precise nature and extent of the differences between reading from either medium have not been completely defined, attempts to identify possible causes of any difference have been made frequently. From a usability perspective, the human is generally considered to respond to stimulation at three levels, physically, perceptually and cognitively. It is not surprising, therefore, that in seeking to understand better the nature of the differences between the media, researchers have sought explanations at these levels. I would add a fourth level to these, the social level, as e-documents have started to take hold in society since the first edition of this book and a host of variables at the organisational and cultural level now seem important to examine if we really seek an understanding of the differences.

In the case of reading it is clear that documents must be handled and manipulated (physical level) before the reader can visually perceive the written information (perceptual level) and thus make sense of the material it contains (cognitive level). Expectations of form and uses for such documents exist among various communities of practice (social level), which serve as a top-down processing guide to the other levels. In a very real sense all these areas are interdependent and therefore the boundaries between the levels are fuzzy rather than rigid. However, 'divide and conquer' is a principle that has long served scientists well and it is to be expected that in seeking to explain media-induced effects in the reading process, researchers have sought to tackle the problem this way.

A complete trawl through the experimental literature at these levels would be exhausting to the reader (not to mention the author) and not best serve our interests here. Furthermore, detailed reviews of large sections of this literature have already been produced (see for example, Dillon 1992). In the present chapter, therefore, a summary of the findings of experimental literature on the subject is outlined in a form that loosely adopts the four-level categorisation outlined earlier.

The emphasis on empirical literature is important. There is no end to the authors who seek to pass opinions or subjective interpretations of the effect of screen presentation on reading. For the most part their opinions are based on very little except their own experiences and if the history of user-centred design has shown nothing else, it has at least demonstrated that the personal experience or opinion of any one advocate is usually of little or no use in explaining complex issues of user psychology.

Physical sources of difference

An electronic text is physically different from a paper one. Consequently, many researchers have examined these aspects of the medium in an attempt to explain the performance differences. Books look different, they afford clues to age, size and subject matter from just the briefest of examinations. They are also highly portable. However, apart from these factors, there are potentially more subtle ergonomic issues involved in reading from screens that some researchers have sought to examine experimentally.

Orientation

One of the advantages of paper over screens is that it can be picked up and oriented to suit the reader. Screens present the reader with only limited flexibility to alter vertical or horizontal orientation, even with laptops. Gould *et al.* (1987a) investigated the hypothesis that differences in orientation may account for differences in reading performance. Users were required to read three articles, one on a vertically positioned screen, one on paper-horizontal and the other on paper-vertical (paper attached via copyholder to equivalent screen). Both paper conditions were read significantly faster than the screen and there were no accuracy differences. While orientation has been shown to affect reading rate of printed material (Tinker 1963) and may have some impact on screen reading too, it does not seem to account for the observed reading differences in the comparisons reported here.

Aspect ratio

The term aspect ratio refers to the relationship of width to height. Typical paper sizes are higher than they are wider, while the opposite is true for typical screen displays. Changing the aspect ratio of a visual field may affect eye-movement patterns sufficiently to account for some of the performance differences. Gould *et al.* (1987a) report a study of 18 subjects reading three eight-page articles on screen, paper and paper-rotated (aspect ratio altered to resemble screen presentation). The results, however, showed little effect for ratio.

It is conceivable, however, that aspect ratio is noted by users more for within-screen formatting than for overall display shape. As such, any screen might allow for the display of text in many different aspect ratios, depending on how the user sizes the document. This would link aspect ratio directly to line length for lengthy texts; for example, a 50-line display with 50 characters per line would have a different aspect ratio than the same text displayed at 20 or 75 characters per line. Manipulating line length directly, Dyson and Kipping (1998) found that lines of 100 characters were read significantly faster than lines of 25 characters, suggesting a benefit to longer lines, which, for any given text may determine aspect ratio. However, further work by Dyson and Haselgrove (2001) indicated that lines this long may incur a comprehension cost, as their data found that when reading for comprehension, a line of 55 characters could be read as fast as a 100 character line and with greater comprehension. Such work continues to indicate how interwoven are many of the physical display variables in reading.

Handling and manipulation: controls, scrolling and paging

Not only are books relatively easy to handle and manipulate, but also one set of handling and manipulation skills serves to operate most document types. Electronic documents, however, cannot be handled directly at all but must be presented through a computer screen and, even then, the operations one can perform and the manner in which they are performed can vary tremendously from one presentation system to another.

Over the past 15 years numerous input devices have been designed and proposed as optimal for computer users; for example, mouse, function keyboard, joystick, light pen, etc. Since Card *et al.*'s (1978) claim that the speed of text selection via a mouse was constrained only by the limits of human information processing, this device has assumed the dominant position in the market. It has since become clear that, depending on the task and users, other input devices can significantly outperform the mouse (Milner 1988). For example, when less than 10 targets are displayed on screen and the cursor can be made to jump from one to the next, cursor keys are faster than a mouse (Shneiderman 1997). In the electronic text domain, Ewing *et al.* (1986) found this to be the case with the HyperTIES application, though in this study the mouse seems to have been used on less than optimal surface conditions.

Although 'direct manipulation' (Shneiderman 1997) might be a common description of an interface, it seems that its current manifestations leave much to be desired when it comes to manipulating text. With the recent popularity of handheld devices, where manipulation is largely handled by small buttons and a stylus, it is clear that users will continue to suffer less than optimal manipulation tools. Obviously, practice and experience will play a considerable part here, but O'Hara and Sellen (1997) showed that even for experienced computer users, the direct manipulation capabilities of paper offer considerable advantages over electronic documents.

It is important to realise that the whole issue of input device cannot be separated from other manipulation variables such as scrolling or paging through documents. For example, a mouse that must be used in conjunction with a menu for paging text will lead to different performance characteristics than one used with a scroll bar. On the basis of a literature review, Mills and Weldon (1986) reported that there is no real performance difference between scrolling and paging electronic documents, though Schwartz *et al.* (1983) found that novices tend to prefer paging (probably based on its close adherence to the book metaphor) and Dillon *et al.* (1990) report that a scrolling mechanism was the most frequently cited improvement suggested by subjects assessing their reading interface, which runs counter to the oft-stated rule that 'users do not scroll'!

Dyson and colleagues have examined scrolling at a more sophisticated level than most in an attempt to identify how it influences comprehension. In a series of studies (see Dyson and Kipping 1998 and Dyson and Haselgrove 2001) they reported that specific scrolling patterns were associated with better comprehension. Dyson and Kipping (1998) show that users can and will scroll and read simultaneously for short but not for long lines.

For the moment, the mouse and scrolling mechanism appears dominant as the physical input and control device for text files, and as the 'point and click' concept they embrace becomes integrated with the 'look and feel' of hypertext they will prove difficult to replace, even if convincing experimental evidence against their use or an innovative credible alternative should emerge.[2]

---

2 One has only to consider the dominance of the far from optimal QWERTY keyboard to understand how powerful convention is. This keyboard was in fact designed to slow human operators down by increasing finger travel distances, thereby lessening key jamming in early mechanical designs. Despite evidence that other layouts can provide faster typing speeds with less errors (Martin 1972), the QWERTY format retains its dominant position.

Display size

Display size is a much discussed but infrequently studied aspect of HCI in general and reading electronic text in particular. Popular wisdom suggests that 'bigger is better' but empirical support for this edict is sparse. Duchnicky and Kolers (1983) investigated the effect of display size on reading constantly scrolling text and reported that there is little to be gained by increasing display size to more than four lines either in terms of reading speed or comprehension. Elkerton and Williges (1984) investigated 1-, 7-, 13- and 19-line displays and reported that there were few speed or accuracy advantages between the displays of seven or more lines. Similarly, Neal and Darnell (1984) reported that there is little advantage in full-page over partial-page displays for text-editing tasks.

These results could be interpreted as suggesting that there is some critical point in display size, probably around five lines, above which improvements are slight. Intuitively this seems implausible. Few readers of paper texts would accept presentations of this format. Experiences with paper suggest that text should be displayed in larger units than this. Furthermore, loss of context is all too likely to occur with lengthy texts and the ability to browse and skim backward and forward is much easier with 30 or so lines of text than with five-line displays (as we discuss later). Richardson *et al.* (1989) report a preference effect for 60-line over 30-line screens but no significant performance effect for an information location-type task in a book-length text. Subsequent work by de Brujin *et al.* (1992) showed that for small screens, users would scroll one line at a time to read, a far from optimal strategy for most reading activities, but certainly evidence of the infinite adaptability of human users.

Conclusion on physical variables

There are wide-ranging physical differences between paper and electronic documents that affect the manner in which readers use them. While myths abound on the theme of display size or orientation, research suggests that the major problems occur in designing appropriate manipulation facilities for the electronic medium. It is interesting to speculate how readers responses may alter as a function of familiarity and emerging standards for manipulation devices. To date, however, the empirical literature would seem to indicate that the major source of the reported performance differences between the media does not lie at the physical level of reading.

Perceptual sources of difference

Most ergonomic work on reading from screens has concentrated on the perceptual differences between screen and paper presented text. With good reason, the image quality hypothesis emerged, which held that many of the reported differences between the media resulted from the poorer quality of image available on screens. An exhaustive programme of work conducted by Gould and his colleagues at IBM between 1982 and 1987 represents probably the most rigorous and determined research effort. They tried to isolate a single variable responsible for observed differences. Reviews and summaries of this work have been published elsewhere (Gould *et al.* 1987a, b; Dillon 1992) and are not reproduced here. However, several issues are worth raising in respect to this hypothesis, of which developers of electronic documents need to be aware.

The image quality hypothesis

For reasons of screen physics, the quality of the image presented to the reader of an electronic document may be variable and unstable. This has led to the search for important determinants of image quality and their relationship to reader performance. Some of the major areas of attention have been flicker, screen dynamics, visual angle of view and image polarity.

Flicker

Since the characters are, in effect, repeatedly fading and being regenerated it is possible that they appear to flicker rather than remain constant. The amount of perceived flicker will obviously depend on both the refresh rate and the phosphor's persistence; the more frequent the refresh rate and the longer the persistence, the less perceived flicker. However, refresh rate and phosphor persistence alone are not sufficient to predict whether or not a user will perceive flicker. It is also necessary to consider the luminance of the screen. While a 30 Hz refresh rate is sufficient to eliminate flicker at low luminance levels, Bauer *et al.* (1983) suggested that a refresh rate of 93 Hz is necessary in order for 99 per cent of users to perceive a display of dark characters on a light background as flicker free, and most contemporary advice is to have a rate of at least 72 Hz for visual comfort.

If flicker was responsible for the large differences between reading from paper and screen it would be expected that studies such as Creed *et al.*'s (1987), which employed photographs of screen displays, would have demonstrated a significant difference between reading from photos and screens. However, the extent to which flicker may have been an important variable in many studies is unknown, as details of screen persistence and refresh rates are often not included in publications. Gould *et al.* (1987a) state that the photographs used in their study were of professional quality but appeared less clear than the actual screen display. It is likely that using photos to control flicker may not be a suitable method and flicker may play some part in explaining the differences between the two media.

Visual angle and viewing distance

Gould and Grischkowsky (1984) first hypothesised that due to the usually longer line lengths on screens the visual angle subtended by lines in each medium differs and that people have learned to compensate for the longer lines on screens by sitting further away from them when reading. They had 18 subjects read 12 different three-page articles for misspellings. Subjects read two articles at each of six visual angles, 6.7, 10.6, 16.0, 24.3, 36.4 and 53.4 degrees, varied by maintaining a constant reading distance while manipulating the image size used. Results showed that visual angle significantly affected speed and accuracy. However, the effects were only noticeable for extreme angles, and between a range of 16.0 to 36.4 degrees, which covers typical screen viewing, no effect for angle was found. Brand and Judd (1993) report that visual angle of hard copy in a text-editing task affected performance in a manner largely consistent with Gould and Grischkowsky (1984). Other work has confirmed that preferred viewing distance for screens is greater than that for paper (Jaschinski-Kruza 1990). Sheih (2000) reported that users who adopted a longer viewing distance from a screen while reading reported less visual fatigue and that reflections on screen tended to draw readers into shorter viewing distances.

Image polarity

A display in which dark characters appear on a light background (for example, black on white) is referred to as positive image polarity or negative contrast. This is referred to here as positive presentation. A display on which light characters appear on a dark background (for example, white on black) is referred to as negative image polarity or positive contrast. This is referred to here as negative presentation. The typical computer display involves positive presentation, typically white on black, though with current web designs, just about any combination is possible!

Since 1980 there has been a succession of publications concerned with the relative merits of negative and positive presentation. Several studies suggest that, tradition notwithstanding, positive presentation may be preferable to negative. For example, Radl (1980) reported increased performance on a data input task for dark characters and Bauer and Cavonius (1980) reported a superiority of dark characters on various measures of typing performance and user preference.

With regards to reading from screens Cushman (1986) reported that reading speed and comprehension on screens was unaffected by polarity, though there was a nonsignificant tendency for faster reading of positive presentation. Gould *et al.* (1987a) specifically investigated the polarity issue. Fifteen subjects read five different 1000-word articles, two negatively presented, two positively presented and one on paper (standard positive presentation). Further experimental control was introduced by fixing the display contrast for one article of each polarity at a contrast ratio of 10 : 1 and allowing the subject to adjust the other article to their own liking. This avoided the possibility that contrast ratios may have been set which favoured one display polarity. Results showed no significant effect for polarity or contrast settings, though 12 of the 15 subjects did read faster from positively presented screens, leading the investigators to conclude that display polarity probably accounted for some of the observed differences in reading from screens and paper.

In a general discussion of display polarity Gould *et al.* (1987b) state that 'to the extent that polarity makes a difference it favours faster reading from dark characters on a light background' (p. 514). Furthermore, they cite Tinker (1963), who reported that polarity interacted with type size and font when reading from paper. The findings of Bauer *et al.* (1983) with respect to flicker certainly indicate how perceived flicker can be related to polarity. Therefore, the contribution of display polarity in reading from screens is probably important through its interactive effects with other display variables.

Display characteristics

Issues related to fonts, such as character size, line spacing and character spacing, have been subjected to detailed research. However, the relationship of much of the findings to reading continuous text from screens is not clear, since some of the tasks employed (for example, Beldie *et al.* 1983) did not include reading continuous text.

Early work on this topic emphasised dot-matrix size (for example, Cakir *et al.* 1980 recommend a minimum of 7×9). Pastoor *et al.* (1983) studied the relative suitability of four different dot-matrix sizes and found reading speed varied considerably. On the basis of these results the authors recommended a 9×13 character size matrix. Recent evidence of developments in resolution, for example, Microsoft's ClearType (Tyrrell *et al.* 2001), show that there is still room for improvement in displays, and preliminary results suggest positive benefits for users.

Muter *et al.* (1982) compared reading speeds for text displayed with proportional or nonproportional spacing and found no effect. In an experiment intended to identify the possible effect of such font characteristics on the performance differences between paper and screen reading, Gould *et al.* (1987a) also found no evidence to support the case for proportionally spaced text.

Kolers *et al.* (1981) studied interline spacing and found that with single spacing, significantly more fixations were required per line, fewer lines were read and the total reading time increased. However, the differences were small and were regarded as not having any practical significance. On the other hand, Kruk and Muter (1984) found that single spacing produced 10.9 per cent slower reading than double spacing, a not inconsiderable difference.

Muter and Mauretto (1991) attempted various 'enhancements' to screen-presented text to see if they could improve reading performance. These included double spacing between lines, proportional spacing within words, left justification only and positive presentation. 'Enhanced' text proved to be read no differently from more typical electronic text (that is, basically similar to paper), which the authors state may be due to one or two of their 'enhancements' having a negative and therefore neutralising effect on others or some 'enhancements' interacting negatively. Unfortunately, their failure to manipulate such variables systematically means firm conclusions cannot be drawn.

Font design has continued to develop with new display technologies and obviously much work needs to be done before a full understanding of the relative advantages and disadvantages of particular formats and types of display is achieved. While Gould *et al.* (1987a) concluded that fonts had little effect, as long as they were reasonably readable, recent developments of screen fonts for web display have shown that there are effects for fonts worth noting. Boyarski *et al.* (1998) studied readability and subjective preference for new binary bitmap fonts designed for electronic presentations, most noticeably for Arial and MS Sans Serif, which were preferred by users and, at key font sizes (9.75), led to improved reading speeds. Once again, the interaction of variables is almost certainly an important determinant of any effect, as acknowledged by Gould *et al.* (1987a) in their original study.

Anti-aliasing

Most computer displays are raster displays typically containing dot matrix characters and lines that give the appearance of 'staircasing', that is, edges of characters may appear jagged. This is caused by undersampling the signal that would be required to produce sharp, continuous characters. The process of anti-aliasing has the effect of perceptually eliminating this phenomenon on raster displays. A technique for anti-aliasing developed by IBM accomplishes this by adding variations in grey level to each character.

The advantage of anti-aliasing lies in the fact that it improves the quality of the image on screen and facilitates the use of fonts more typical of those found on printed paper. To date, the only reported investigation of the effects of this technique on reading from screens is that of Gould *et al.* (1987a). They had 15 subjects read three different 1000-word articles, one on paper, one on screen with anti-aliased characters and one on screen without anti-aliased characters. Results indicated that reading from anti-aliased characters did not differ significantly from either paper or aliased characters, though the latter two differed significantly from each other. Although the trend was present the results were not conclusive and no certain evidence for the effect of anti-aliasing was provided. However, the authors report that 14 of the 15 subjects preferred the anti-aliased characters, describing them as clearer and easier to read.

The interaction of display variables: the image quality hypothesis

Despite many of the findings reported thus far, it appears that reading from screens can be at least as fast and as accurate as reading from paper. Gould *et al.* (1987b) have demonstrated empirically that under the right conditions and for a strictly controlled proofreading task, such differences between the two presentation media cease to be significant. In a study employing 16 subjects, an attempt was made to produce a screen image that closely resembled the paper image, that is, similar font, size, colouring, polarity and layout were used. Univers-65 font was positively presented on a monochrome IBM 5080 display with an addressability of 1024×1024. No significant differences were observed between paper and screen reading. This study was replicated with 12 further subjects using a 5080 display with an improved refresh rate (60 Hz). Again no significant differences were observed though several subjects still reported some perception of flicker, probably due to the refresh rate, which would be considered low by today's standards.

On balance it appears that any explanation of these results must be based on the interactive effects of several of the variables outlined in the previous sections. After a series of experimental manipulations aimed at identifying those variables responsible for the improved performance, Gould *et al.* (1987b) suggested that the performance deficit was the product of an interaction between a number of individually nonsignificant effects. Specifically, they identified display polarity (dark characters on a light, whitish background), improved display resolution, and anti-aliasing as major contributions to the elimination of the paper/screen reading rate difference.

Gould *et al.* (1987b) conclude that the explanation of many of the reported differences between the media is basically visual rather than cognitive and lies in the fact that reading requires discrimination of characters and words from a background. The better the image quality is, the more reading from screen resembles reading from paper and hence the performance differences disappear. This seems an intuitively sensible conclusion to draw. It reduces to the level of simplistic any claims that one or other variable such as critical flicker frequency, font or polarity is responsible for any differences. As technology improves we can expect to see fewer speed deficits at least for reading from screens. Furthermore, evidence from Muter and Maurutto (1991) using a commercially available screen has shown this to be the case, although other differences remain.

A major shortcoming of the studies by Gould *et al.* is that they only address limited outcome variables: speed and accuracy. Obviously speed is not always a relevant criterion in assessing the output of a reading task. Furthermore, the accuracy measures taken in these studies have been criticised as too limited and further work needs to be carried out to appreciate the extent to which the explanation offered by Gould is sufficient. It follows that other observed outcome differences such as fatigue, reader preference and comprehension should also be subjected to investigation in order to understand how far the image quality hypothesis can be pushed as an explanation for reading differences between the two media.

A shortcoming of most work cited in this section is the task employed. Invariably it was proofreading, which hardly constitutes normal reading for most people. Thus the ecological validity of many of these studies is low. Beyond this, the actual texts employed were all relatively short (Gould's, for example, averaged only 1100 words and many other researchers used even shorter texts). As a result, it is difficult to generalise these conclusions beyond the specifics of task and texts employed to the wider class of activities termed 'reading'. Creed *et al.* (1987) defend the use of proofreading on the grounds of its amenability to manipulation and control. While this desire for experimental rigour is laudable, one cannot but feel that the major issues involved in using screens for real-world reading scenarios are not addressed by such work.

Cognitive sources of difference

It is clear that the search for the specific ergonomic variables responsible for differences between the media has been insightful. However, few readers of electronic texts would be satisfied with the statement that the differences between the media are physical or visual rather than cognitive. This might explain absolute speed and accuracy differences on limited tasks but hardly accounts for the range of process differences that are observed or experienced. Once the document becomes too large to display on a single screen, other factors than image quality immediately come into play. At this stage readers must start to manipulate the document and thus be able to relate current to previously displayed material. In such a situation other factors such as memory for text and its location, ability to search for items and speed of movement through the document gain importance and the case for image quality as the major determinant of performance is less easy to sustain. In this section, other possible factors at the cognitive level are discussed.

Short-term memory for text

A related issue to display size and scrolling/paging mentioned earlier is the splitting of paragraphs midsentence across successive screens. In this case, which is more likely to occur in small displays, the reader must manipulate the document to complete the sentence. This is not a major issue for paper texts such as books or journals because the reader is usually presented with two pages at a time and access to previous pages is normally easy. On screen, however, access rates are not so fast and the break between screens of text is likely to be more critical.

Research into reading has clearly demonstrated the complexity of the cognitive processing that occurs. The reader does not simply scan and recognise every letter in order to extract the meaning of words and then sentences. Comprehension is thought to require inference and deduction, and the skilled reader probably achieves much of his/her smoothness by predicting probable word sequences (Chapman and Hoffman 1977, but see Mitchell 1982). The basic units of comprehension in reading that have been proposed are propositions (Kintsch 1974), sentences (Just and Carpenter 1980) and paragraphs (Mandler and Johnson 1977). Splitting sentences across screens is likely to disrupt the process of comprehension by placing an extra burden on the limited capacity of working memory to hold the sense of the current conceptual unit while the screen is filled. Furthermore, the fact that 10-20 per cent of eye movements in reading are regressions to earlier fixated words and that significant eye-movement pauses occur at sentence ends (Ellis 1983) would suggest that sentence splitting is also likely to disrupt the reading process and thereby hinder comprehension.

Dillon *et al.* (1990) found that splitting text across screens caused readers to return to the previous page to reread text significantly more often than when text was not split, a finding replicated by de Brujin *et al.* (1992). Although this appeared to have no effect on subsequent comprehension of the material being read, they concluded that it was remarked upon by the subjects sufficiently often to suggest that it would be a nuisance to regular users. In this study, however, the subjects were reading from a paging rather than scrolling interface where the effect of text splitting was more likely to cause problems due to screen-fill delays. With scrolling interfaces text is always going to split across screen boundaries but there is rarely a perceptible delay in image presentation to disrupt the reader. It would seem, therefore, that to the extent to which such effects are likely to be noticeable, text splitting should be avoided for paging interfaces.

Visual memory for location

There is evidence to suggest that readers establish a visual memory for the location of items within a printed text based on their spatial location both on the page and within the document (Rothkopf 1971; Lovelace and Southall 1983). This memory is supported by the fixed relationship between an item and its position on a given page. A scrolling facility is therefore liable to weaken these relationships and offers the reader only the relative positional cues that an item has with its immediate neighbours.

It has become increasingly common to present information on computer screen via windows, that is, sections of screen devoted to specific groupings of material. Current technology supports the provision of independent processes within windows or the linking of inputs in one window with the subsequent display in another, the so-called 'coordinated windows' approach (Shneiderman 1997).

Such techniques have implications for the presentation of text on screen as they provide alternatives to the straightforward listing of material in 'scroll' form or as a set of 'pages'. For example, while one window might present a list of contents in an electronic text, another might display whole sections of it according to the selection made. In this way, not only is speed of manipulation increased but also the reader can be provided with an overview of the document's structure to aid orientation while reading an opened section.

Tombaugh *et al.* (1987) investigated the value of windowing for readers of lengthy electronic texts. They had subjects read two texts on single or multiwindow formats before performing 10 information location tasks. They found that novices initially performed better with a single-window format but subsequently observed that, once familiar with the manipulation facilities, the benefits of multiwindowing in terms of aiding spatial memory became apparent. They highlight the importance of readers acquiring familiarity with a system and the concept of the electronic book to accrue the benefits of such facilities.

Simpson (1989) compared performance with a similar multiwindow display, a 'tiled' display (in which the contents of each window were permanently visible) and a 'conventional' stack of windows (in which the windows remained in reverse order of opening). She reported that performance with the conventional window stack was poorest but that there was no significant difference between the 'tiled' and multiwindow displays. She concluded that for information location tasks, the ability to see a window's contents is not as important as being able to identify a permanent location for a section of text.

Stark (1990) asked people to examine a hypertext document in order to identify appropriate information for an imaginary client and manipulated the scenario so that readers had to access information presented either in a 'pop-up' window, which appeared in the top right-hand corner of the screen, or in a 'replacement' window, which overlaid the information currently being read. Although no significant task performance or navigation effects were observed, subjects seemed more satisfied with pop-ups than replacements.

Such studies highlight the impact of display format on readers' performance of a standard reading task: information location. Spatial memory seems important and paper texts are good at supporting its use through permanence of format. Windowing, if deployed so as to retain order, can be a useful means of overcoming this inherent weakness of electronic text. However, studies examining the problems of windowing very long texts (where more than five or six stacked windows or more frequent window manipulations are required) need to be performed before any firm conclusions about the benefits of this technique can be drawn. If anything, such thoughts lend support to the general use of frames, even though web designers usually dismiss this facility outright.

Schematic representations of documents

Exposure to the variety of texts in everyday life could lead readers to develop mental models or schemata for documents with which they are familiar. Schemata are hypothetical knowledge structures that mentally represent the general attributes of a concept and are considered by psychologists to provide humans with powerful organising principles for information (Bartlett 1932; Cohen 1988). Thus, when we pick up a book, we immediately have expectations about the likely contents. Inside the front cover we expect such details as where and when it was published, perhaps a dedication and then a Contents page. We know, for example, that contents listings describe the layout of the book in terms of chapters, proceeding from the front to the back. Chapters are organised around themes and an index at the back of the book, organised alphabetically, provides more specific information on where information is located in the body of the text. Experienced readers know all this before even opening the text. It would strike us as odd if such structures were absent or their positions within the text were altered.

According to van Dijk and Kintsch's (1983) theory of discourse processing, such models or schemata, which they term 'superstructures', facilitate comprehension of material by allowing readers to predict the likely ordering and grouping of constituent elements of a body of text. To quote van Dijk (1980): 'a superstructure is the schematic form that organises the global meaning of a text. We assume that such a superstructure consists of functional categories . . . (and) . . . rules that specify which category may follow or combine with what other categories' (p. 108).

But apart from categories and functional rules, van Dijk adds that a superstructure must be socioculturally accepted, learned, used and commented upon by most adult language users of a speech community. Research by van Dijk and Kintsch (1983) and Kintsch and Yarborough (1982) has shown how such structures influence comprehension of texts.

In this format the schema/superstructure constitutes a set of expectancies about their usual contents and how they are grouped and positioned relative to each other. In advance of actually reading the text, readers cannot have much insight into anything more specific than this, but the generality of organisation within the multitude of texts read in everyday life affords stability and orientation in what could otherwise be a complex informational environment.

The concept of a schema for an electronic information space is less clear-cut than those for paper documents. Electronic documents have a far shorter history than paper and the level of awareness of technology among the general public is relatively primitive compared to that of paper. Exposure over time will almost certainly improve this state of affairs but even among the contemporary computer literate, generic schematic structures for information on paper have few electronic equivalents of sufficient generality.

Obviously, computing technology's short history is one of the reasons, but it is also the case that the medium's underlying structures do not have equivalent transparency. Thus using electronic information is often likely to involve the employment of schemata for systems in general (that is, how to operate them) in a way that is not essential for paper-based information.

The qualitative differences between the schemata for paper and electronic documents can easily be appreciated by considering what you can tell about either at first glance. The information available to paper text users was outlined earlier. When we open a hypertext or other electronic document, however, we do not have the same amount of information available to us. We are likely to be faced with a welcoming screen that might give us a rough idea of the contents (that is, subject matter) and information about the authors/developers of the document, but little else. Such displays are usually two-dimensional, give no indication of size, quality of contents, age (unless explicitly stated) or how frequently the text has been used (that is, there is no dust or signs of wear and tear on it such as grubby finger-marks or underlines and scribbled comments).

Performing the electronic equivalent of opening up the text or turning the page offers no assurance that expectations will be met. Many hypertext documents offer unique structures (intentionally or otherwise) and their overall sizes are often impossible to assess in a meaningful manner (these points are dealt with in more detail in Dillon *et al.* 1993). At their current stage of development it is likely that users/readers familiar with hypertext have a schema that includes such attributes as linked nodes of information, nonserial structures and, perhaps, potential navigational difficulties! The manipulation facilities and access mechanisms available in hypertext will probably occupy a more prominent role in their schemata for hypertext documents than they will for readers' schemata of paper texts. As yet, empirical evidence for such schemata is lacking though it is emerging (Dillon and Gushrowski 2000).

The fact that hypertext offers authors the chance to create numerous structures out of the same information is a further source of difficulty for users or readers. Since schemata are generic abstractions representing typicality in entities or events, the increased variance of hypertext implies that any similarities that are perceived must be at a higher level or must be more numerous than the schemata that exist for paper texts.

It seems, therefore, that users' schemata of electronic texts are likely to be 'informationally leaner' than those for paper documents. This is attributable to the recent emergence of electronic documents and comparative lack of experience interacting with them as opposed to paper texts for even the most dedicated users. The lack of standards in the electronic domain compared to the rather traditional structures of many paper documents and the paucity of affordances or cues to use are further problems for schema development with contemporary electronic texts.

Searching

Electronic text supports word or term searches at rapid speed and with total accuracy and this is clearly an advantage for users in many reading scenarios; for example, checking references, seeking relevant sections, etc. Indeed it is possible for such facilities to support tasks that would place unreasonable demands on users of paper texts; for example, searching a large book for a nonindexed term or several volumes of journals for references to a concept.

Typical search facilities require the user to input a search string and choose several criteria for the search, such as ignoring certain text forms (for example, all uppercase words), but sophisticated facilities on some database systems can support specification of a range of texts to search. The usual form for search specification is Boolean, that is, users must input search criteria according to formal rules of logic employing the constructs 'either', 'or' as well as 'and', which when used in combination support powerful and precise specifications. Unfortunately, most end-users of computer systems are not trained in their use, and while the terms may appear intuitive, they are often difficult to employ successfully. A barrage of empirical evidence now exists on user difficulties in searching generally (see for example, Pollock and Hockley 1997) and most users are slow to use Boolean operators, even if they understand them.

In current electronic text facilities a simple word search is common but users still seem to have difficulties. Richardson *et al.* (1988) reported that several subjects in their experiment displayed a tendency to respond to unsuccessful searches by increasing the specificity of the search string rather than lessening it. The logic appeared to be that the computer required precision rather than approximation to search effectively. While it is likely that such behaviour is reduced with increased experience of computerised searching, a study by McKnight *et al.* (1989) of information location within text found other problems. Here, when searching for the term 'wormwood' in an article on winemaking, two subjects input the search term 'woodworm', displaying the intrusion of a common-sense term for an unusual word of similar sound and shape (a not uncommon error in reading under pressure due to the predictive nature of this act during sentence processing). When the system correctly returned a 'Not Found' message, both users concluded that the system could not fail, hence the question was an experimental trick.

Thus it seems as if search facilities are a powerful means of manipulating and locating information on screen and convey certain advantages impossible to provide in the paper medium. However, users may have difficulties with them in terms of formulating accurate search criteria. This is an area where research into the design of search facilities and increased exposure of users to electronic information can lead to improvements resulting in a positive advantage of electronic text over paper. Kelly and Cool (2002), perhaps unsurprisingly, showed that increased familiarity with material led to more efficient searching.

Individual differences in skill, knowledge and experience

It has been noted that many of the studies reported in this review employed relatively naive users as subjects. The fact that different types of users interact with computer systems in different ways has long been recognised and it is possible that the differences in reading that have been observed in these studies result from particular characteristics of the user group involved.

Most obviously, it might be assumed that increased experience in reading from computers would reduce the performance deficits. A direct comparison of experienced and inexperienced users was incorporated into a study on proofreading from screens by Gould *et al.* (1987a). Experienced users were described as 'heavy, daily users . . . and had been so for years'. Inexperienced users had no experience of reading from computers. No significant differences were found between these groups, both reading slower from screen.

While experience alone might not be significant, Dyson and Haselgrove (2001) have shown that training can impact performance when reading from screen. In their study, users who were trained to read faster managed to improve their reading speeds for long (100-character) lines, again showing that user characteristics are dynamic rather than fixed.

Smedshammar *et al.* (1989) report that post-hoc analysis of their data indicates that fast readers are more adversely affected by screen presentation than slow readers. However, their classification of reading speed is based on mean performance over three conditions in their experiment rather than controlled, pretrial selection, suggesting caution in drawing conclusions. Smith and Savory (1989) report an interaction effect between presentation medium, reading strategy and susceptibility to external stress measured by questionnaire, suggesting that working with screens may exaggerate some differences in reading strategy for individuals with high stress levels. Caution in interpretation of these results is suggested by the authors.

No reported differences for age or sex can be found in the literature. Therefore, it seems reasonable to conclude that basic characteristics of the user are not responsible for the differences in reading from these presentation media.

Social differences

While the other levels of analysis are well represented in the literature on reading from screens, there have been few empirical analyses of the use of paper and screens that have adopted a purely social level of analysis. Some of the work at the cognitive level has hinted at the cultural forces shaping human responses to documents (the superstructure work of van Dijk and Kintsch, for example).

Borgman (2000) is keen to draw parallels between what is happening today with digital libraries and electronic documents and what Guttenberg launched upon the world. Strong on history, she makes clear that even that revolution was built on the antecedent work of others who wrestled with similar ideas. Similarly, while all eyes now focus on the World Wide Web, Borgman notes that the convergence of tasks and technologies over the past 25 years has led us to the place we now find ourselves in. An extension of this line of analysis is the implication that digital documents, and the technologies that we use to exploit them, are too immature yet for us as a community to have learned to exploit them.

Violating many of the rules we have come to expect in the paper world, e-text may just carry with it an overhead in use (as seen in some of the perceptual and cognitive comparisons) that will be overcome with time. This is a popular argument though it is difficult to see how we can tackle this empirically, not least when the data we have from experienced computer users (for example, Gould *et al.* 1987a; O'Hara and Sellen 1997) suggest that differences between the media do not completely disappear when users are more experienced.

Although it does not point to explicit differences between the media, work by Dillon and Gushrowski (2000) showed that the Web is giving rise to new forms of information genre that have no equivalents in the paper domain. They point out that even when the form is uniquely Web-based, users develop expectations of form and content that are shared among producers and consumers. Reiffel (1999) argues that various communities of users have evolved specific communication forms that are probably not well supported by existing electronic instantiations and that we will only discover what works best in digital space over time. The results may prove very different from the genres we employ in the paper world. While the experimental literature is sparse in this area, it would seem a fruitful area for further enquiry even if we cannot attribute many of the differences reported in this chapter purely to social level factors. Given the trends we see in the adoption of computers for reading on college campuses (Jones 2002), such as the expectation of ubiquitous access, a willingness to download and share files, and a major usage shift from paper and libraries to digital documents, we are perhaps at the dawn of mass acceptance that has eluded us with previous generations.

General conclusion: so what do we know now?

At the outset it was stated that reading can be assessed in terms of outcome and process measures. To date, however, most experimental work has concentrated on the former and, in particular, has been driven by a desire to identify a single variable to account for the significant reading speed differences that have been reported. The present review sought to examine the experimental literature with a view to identifying all relevant issues and show how single variable explanations are unlikely to offer a satisfactory answer.

While substantial progress has been made in terms of understanding the impact of image quality on reading speed, it is clear that ergonomists are still a long way from understanding fully the effect of presentation medium on reading. While it is now possible to draw up recommendations on how to ensure no speed deficit for proofreading short texts on screen, changes in task and text parameters mean such advice has less relevance.

One is struck in reviewing this literature by the rather limited view of reading that some investigators seem to have. Most seem to concern themselves with the control of so many variables that the resulting experimental task bears little resemblance to the activities most of us routinely perform under the banner 'reading'. It is perhaps no coincidence that the major stumbling block of reader preference has been so poorly investigated beyond the quick rating of screens and test documents in postexperimental surveys.

The assumption that overcoming speed or accuracy differences in proofreading is sufficient to claim, as some authors have, that 'there is no difference' between the media (Oborne and Holton 1988) is testimony to the limitations of some ergonomists' views of human activities such as reading. Other tasks, such as reading to comprehend, to learn or for entertainment, are less likely to require readers to concern themselves with speed. These are the sort of tasks people will regularly wish to perform and it is important to know how electronic text can be designed to support them. Such tasks will also of necessity involve a wide variety of texts, differing in length, detail, content-type and so forth – issues that have barely been touched upon to date by researchers.

The findings on image quality and the emerging knowledge of manipulation problems should not be played down, however. Knowing what makes for efficient visual processing and control of electronic text can serve as a basis for future applications. As Muter and Maurutto (1991) demonstrated, a typical high-quality screen with effective manipulation facilities can provide an environment that holds its own in speed, comprehension and preference terms with paper, at least over the relatively constrained reading scenarios found in the researchers' laboratory. But if our desire is to create systems that improve on paper rather than just matching it in performance and satisfaction terms (as it should be), then much more work and a more realistic conceptualisation of human reading is required.

4 Describing the reading process at an appropriate level

Some books are to be tasted, others to be swallowed, and some few to be chewed and digested: that is, some books are to be read only in parts, others to be read, but not curiously, and some few to be read wholly, and with diligence and attention.

Sir Francis Bacon, 1621

Introduction

Taken in isolation, the empirical literature on reading from screens versus paper is often not directly helpful to a designer or design team concerned with developing an electronic text application. The classification of the empirical literature provided in the previous chapter is an attempt at better conceptualising the relevant issues and experimental findings but it suffers from the problems inherent in the literature itself: the absence of a suitable descriptive framework of the reader or reading process that would enable those concerned with electronic text to derive guidance for specific design applications.

As it stands, the empirical literature presents two implicit views of the typical reader and provides recommendations accordingly. The first is as a scanner of short texts, searching out spelling mistakes or some relatively trivial error, often in a race against the clock (and usually trying to impress an experimenter). The second is as a navigator travelling through a maze of information in search of a target. These extremes are tempered only slightly by concessions to text or task variables as influences on the reading process and it is rare that any attempt to place reading in a broader, more realistic context is made. Yet reading is a form of information usage that rarely occurs as a self-contained process with goals expressible in terms of speed or number of items located. Far more frequently it occurs as a means to, or in support of, nontrivial everyday ends such as keeping informed of developments at work, checking bank balances, understanding how something works (or more likely, why right now it isn't working) and so forth. These are precisely the type of information acts, of which reading is an essential component, which people perform routinely. The success of these acts is measured in effects not usually quantifiable simply in terms of time taken or errors made. True, readers do notice spelling mistakes, they may even proofread as professionals and they certainly must manipulate and navigate through lengthy documents, but such views alone can never adequately describe the situated realities and totality of the reading situation.

This constrained view of the reading process becomes even more apparent when one examines the studies of reading that dominate the various disciplines which lay claim to some interest in it. Psychology, a discipline that might justifiably consider itself directly concerned with understanding reading is, according to Samuels and Kamil (1984), concerned with 'the *entire* process from the time the eye meets the page until the reader experiences the "click of comprehension"' (p. 185, italics added). This sounds suitably all-embracing but in reality is relatively narrow when one realises the everyday attributes of reading that it overlooks.

There are few psychological models of reading that consider text manipulation or navigation, for example, as part of the reading process. The literature that provides theoretical input to these domains is usually the product of other research into issues such as memory organisation or learning. Can any view that claims reading only starts when the 'eye meets the page' really lay claim to covering the 'entire' process? Certainly it appears logical to start here, but if the situation demands that the reader moves from text to text in their search for information, do we conclude that the intervening moves are not part of the process of reading? And if readers fail to experience that 'click of comprehension', has the reading process stopped at some point when the eye leaves the text or does it extend while they think the material through and get that 'click' later? Interestingly, not all psychologists would even accept Samuels and Kamil's definition of psychology's legitimate concerns with reading. Crowder and Wagner (1992), for example, explicitly exclude comprehension as an issue in their analysis, where they define reading as ending 'more or less where comprehension begins' (p. 4).

Rumelhart's (1977) theory of reading marked a supposed breakthrough in cognitive models of reading by highlighting the limits of linear models (those proposing a one-way sequence of processing stages from visual input to comprehension). He outlined an alternative interactive form (one supporting the influence of higher stages of processing on lower ones) that accounted for some of the experimental findings that were difficult to accommodate in linear models. His model, parts of which have been successfully implemented in software form thereby passing possibly the strongest test of a contemporary cognitive theory (McClelland and Rumelhart 1981), represents reading largely as an act of word recognition, and has been summarised as follows:

> Reading begins with the recognition of visual features in letter arrays. A short lived iconic image is created in brief sensory storage and scanned for critical determinants. Available features are fed as oriented line

segments into a pattern synthesizer that, as soon as it is confident about what image has been detected, outputs an abstract characterisation ... The extracted features are constraints rather than determiners, interacting with context and reader expectations ... The individual letters are heavily anticipated by stored representations in a 'word index'. Even in recognition of letters and words all of the various sources of knowledge, both sensory and non-sensory, come together in one place and the reading process is the product of the simultaneous joint application of all knowledge sources.

(de Beaugrande 1981, p. 281)

From such a description it is not difficult to understand why Venezky (1984) states that: 'the history of research on the reading process is for the most part the history of cognitive psychology' (p. 4), and, one might add, the reason why cognitive psychology is often accused of an undercritical acceptance of the computer metaphor of mind. However, such conceptualisations, no matter how justifiable theoretically, suffer serious limitations when applied to the work of system designers. Typical cognitive theories of reading fail to mention a reading task, a text, a goal, or a context in which these processes occur. 'Visual feature recognition', 'iconic images' and 'pattern synthesizers' are theoretical constructs which attempt to provide a plausible account of how humans extract information from text (or 'letter arrays' to use the jargon) but mapping findings from such analyses and models to the world of electronic text design is, by definition, beyond the model's scope. If it is usability you are interested in, you had better look elsewhere!

The narrow focus of much reading research is reflected sharply in the opening pages of psycholinguist Frank Smith's (1978) book entitled *Reading*, where he remarks that a glance though the text might leave one justifiably thinking that 'despite its title, this book contains little that is specifically about reading' (p. 1).

Typical arguments in favour of this narrow perspective argue for the need to examine reading in a reductionist manner (claiming that there is little about reading that is unique – it involves certain cognitive processes and structures that researchers not interested in reading have already investigated in other contexts – presuming that such work transfers appropriately to discussions of the reading process, which is itself a debatable assumption), but it is Smith's early admission of the lack of real-life relevance of the work that produced the most memorable 'click of comprehension' in this reader's mind. This is no historical oddity, a function merely of the focus in the 1970s and 1980s on cognitive models. The problems remain, 20 years later, as can be seen in Underwood and Batt's (1996) text entitled *Reading and Understanding*, wherein they state:

The closest we can come to defining reading is by use of a generality, by suggesting that it is a form of problem solving that is directed at the integration of words in an attempt to recover the writer's ideas. By describing reading as 'problem solving' we have, of course, used one mysterious activity to describe another, and we present ourselves the difficulty of saying what it means to solve a problem. Reading, like all forms of problems solving, can be described as an information processing task.

(p. 2)

The act of reading has obviously attracted much attention from cognitivists but this attention has not always yielded practical insights. In a book given over to attacking much of contemporary experimental psychology, Kline (1988) singles out reading as a prime example of the lack of validity in much of the discipline's work. Describing a typical reading experiment investigating people's categorisation of sentences as meaningful or nonmeaningful while their reaction time is measured, he states:

the task . . . is not really like reading. Certainly it involves reading but most people read for pleasure or to gain information. Furthermore, reading has serious emotional connotations on occasion, as for example reading pornography, a letter informing you that your PhD has been turned down or your lover is pregnant . . . Furthermore, most adults, when reading books especially, read large chunks at a time.

(p. 36)

Kline continues, humorously comparing lines from a Shakespeare sonnet (for example, 'Like as the waves make towards the pebbled shore, so do our minutes hasten to their end') with lines from such experimental tasks (for example, 'Canaries have wings – true or false?'; 'Canaries have gills – true or false?'), before concluding that such work is absurd in the context of real reading and its resultant theoretical models of no predictive and little explanatory use. His criticisms might seem harsh and populist were it not for the fact that Kline is a psychologist of international reputation and admits to having performed, like most psychologists, such experiments himself earlier in his career.

But psychology is not unique in failing to provide a satisfactory account of the process for the purposes of design. Information science, the contemporary theoretical backbone of librarianship, might also be viewed as having a natural interest in the reading process. Yet its literature offers few clues to those concerned with designing electronic texts for reading. As Hatt (1976) puts it:

A great body of professional expertise has been developed, the general aims of which have been to improve the provision of books and to facilitate readers' access to books. At the point where the reader and the book come together however, it has been the librarian's habit to leave the happy pair and tiptoe quietly away, like a Victorian novelist.

(p. 3)

Again, the situation has not improved in the intervening years. Julien and Duggan (2000) reviewed the general literature in library and information science on information uses and concluded that much of it is atheoretical. While Pettigrew and McKechnie (2001) argued that theoretical analysis of information use was improving, none of their listed theories in contemporary information science contained any analysis of reading that could help us in this context. But the problems of information science are the problems of all disciplines concerned with this subject and although much valuable work has been done and knowledge has been gained, one comes away from the literature thinking 'that's all well and good, but it's not really reading!'

In defense of each discipline it must be said that their theories, approaches and methods reflect their aims. If psychology really is concerned with what happens between the moments when the eye meets the page and the reader understands the text (or just before that in the case of Crowder and Wagner, among others), then models of eye movements and word recognition have a place, despite Kline's enthusiastic dismissal. One might add that if this is all that many psychologists consider important in understanding the reading process, then Kline might really have a point. Few, if any, theorists interested in reading claim to cover all issues. What is pertinent here, however, is the irrelevance of much of this work to the issues associated with the development of more usable electronic documents, the depressing fact that much excellent work in psychology and allied sciences fails to provide much by way of influence in practical software design settings.

The unsuitability of any theoretical description of reading is a major (perhaps even *the*) problem for human factors work. Viewing the reader as an 'information processor' or an 'information seeker' and the reading process as a 'psycholinguistic guessing game' depending on theoretical stance, hardly affords prescriptive measures for the design of electronic text systems. The typical reader can certainly be said to process information and occasionally use libraries, but each is only a small part of the whole that is reading. If one deals exclusively with such limited aspects (as many theories do), the broad picture never emerges and this gives rise to the type of limited findings on text design one finds in much of the human factors literature.

The problem of theoretical description for human factors work

Since the first edition of this book there has been a growing recognition of the problems inherent in HCI's unquestioned borrowing of existing theoretical orientations from related disciplines, but relevant theory has been slow to emerge. User-advocates are strong on empirical rather than theoretical methods and the greatest effort here is placed on gaining user inputs to the process so as to guide emerging designs.

The problem for human factors work induced by our lack of theoretical descriptions is epitomised in a case study of a commercial system involving the author (Dillon *et al.* 1991). A publishing consortium funded the development of an experimental system to support the document supply industry. Named ADONIS, the resulting workstation was designed to facilitate searching, viewing and printing of articles. It boasted a high-resolution A4-size screen that presented bit-mapped reproductions of journal articles. The trial system presented users with access to biomedical journals (selected on the basis of a usage study) and the workstation came to be seen as a prototype electronic text system of the future.

The author was asked to evaluate the system for both user types and conducted a cognitive walkthrough, and the setting of three tasks for a sample of 10 users to perform in an informal manner, that is, with the evaluator present and the participants commenting on the system's user interface as they worked through the tasks. Measures of speed and accuracy were eschewed in favour of general ratings of the system and comments on good or bad aspects of the interface, which the evaluator noted as the subjects proceeded. As a result of frequent criticisms of the search facilities, a survey of normal procedures for citing articles among 35 researchers was also carried out to identify a more suitable set and layout for the search fields. The results of the evaluations were summarised, related to the literature on electronic text and general interface design and presented to the publishing consortium in the form of a written report.

Superficially, ADONIS was a good design. The large, high-quality screen presented articles in an easy-to-read manner that conformed precisely to the structure of the paper version. By using a large screen and positive presentation it even adhered to some of the human factors design guidelines in the literature. Use of menus and a form-filling screen for inputting search criteria should have removed any learning difficulties for novice users too. The ability to store and retrieve a large number of articles from one system, coupled with the ability to view material on screen before deciding whether or not to print it out, seemed to convey benefits to the ADONIS workstation not even possessed by paper.

The users, however, were very critical of the system. Common criticisms related to the style of the interface, which was described as 'archaic', the speed of searching, which was perceived as too slow, the inability to search on keywords, and the restricted manipulation facilities available when an article was being viewed. Only two users said they would read with it; of the remainder, six said they would only scan articles prior to printing them out and two said they would never use it. In other words, although the system was designed in partial accordance with the literature on electronic text, these potential users rejected it. How could this be?

What was shown by the evaluation is that while the system supported the ends, it failed to adequately provide the means. In other words, it emphasised functionality at the expense of usability, letting users get *what* they want but not *how* they want to get it. Users could obtain hardcopies of journal articles but they had to master the counterintuitive specification form first.[1] They could browse articles on a high-quality large screen, but they could not manipulate pages with the ease they can with paper. They could search an equivalent of a library of journals from their desk to obtain an article they seek, but they could not browse through a list of contents and serendipitously discover a relevant title or author as they can with paper.

The levels of description problem in user experience design

This clash between means and ends provides an interesting insight into the problems faced by many designers of electronic text systems (or indeed, information systems in general), which will be referred to here as the 'levels of description' problem. Briefly stated, it implies that there are various levels of abstraction at which human behaviour can be described (for example, the physical, the cognitive, or the social level) and, while each may be accurate in itself, there exists an optimum level for any given use (for example, analysing consumer spending requires different views of human activity than describing human task performance when driving). In the case of systems design, using a nonoptimum level leads to superficial matching of design to needs if the level is too shallow, and to an inability to specify designs for needs if the level is too deep. These are elaborated with two examples pertinent to electronic text design.

ADONIS seemed to match basic reader requirements. However, it was obvious that it did so only at a superficial level. By describing readers' needs at the gross level of 'obtaining a hardcopy', 'locating an article', 'browsing the references' and so forth, it made (and matched) design targets of accurate but inadequately specified needs. The designers obviously developed a product that meets these targets, but only at the grossest level of described behaviour. A description of reading at a deeper level than this might well have produced a different set of requirements and resulted in a more usable design.

An example of a level of description too deep to specify needs for design can be found in most of the work on modelling reading by cognitive psychologists. By concentrating on theoretical structures and processes in the

---

1 The survey of citation style revealed that users tend to refer to articles in the form author/year/title/journal, or author/journal/article/year. ADONIS structured input in the form: ADONIS number/ISSN Number/journal/year/author, etc., which was considered very confusing by some users and led to frequent errors during trial tasks.

reader's mind or eye movements in sentence perception, word recognition and so forth, such work aims to build a body of knowledge on the mental activities of the reader. Fascinating as this is, it is difficult to translate work from this level of description or analysis to the specification of software intended to support the reading process. For a clear example of this, just attempt to draw a set of guidelines from any theory of reading that is applicable to HCI.

Highlighting limitations is only useful where it serves to advance the process of overcoming them. What is required is a description or form of discourse that bridges between these levels and supports valid descriptions of human activities in a form that is most meaningful for system design. This is not an easy task, but one is helped by at least knowing where the goal-posts are. Within the electronic text domain a suitable analytic framework should provide designers with a means of posing appropriate questions and deriving relevant answers. Clearly existing ones, be they psychological, typographical or information scientific, do not. How then should we conceptualise the reading process?

Identifying an appropriate abstraction: the vertical slice

It is unlikely that the evolution of a suitable description of the reading process will result merely from performing more experiments on reading from screens. To attempt empirical testing of all conceivable reading scenarios would be impossible and even the application of demonstrable ergonomic principles derived from such work (for example, the importance of image quality) is insufficient to guarantee successful design.

In the previous chapter I categorised the literature on reading from screens into four levels: the physical, perceptual, cognitive and social. Most research exists at only one of these levels and while there is good reason to constrain one's research question to a single level through the human condition, we require what I can best describe as a 'vertical slice' through all of them. In other words, we need a model or theory that portrays readers as multilevelled beings with needs and issues at each level.

Rasmussen (1986) describes the need for a multilayered analysis involving descriptions ranging from the social function of a system, through the information processing capabilities of the user and machine, to the physical mechanisms and anatomy of both the user and the machinery. He emphasises the need to incorporate perspectives of human abilities from quite separate research paradigms in order to describe usefully the process of interaction with advanced technology and adds: 'it is important to identify manageable categories of human information processes at a level that is independent of the underlying psychological mechanisms' (p. 99).

Put another way, the framework needed for design should not be overly concerned with the architecture of human cognition (as, by definition, is the case with cognitive psychological models of reading) but should address primarily the nature of the processing activity. Thus, according to Rasmussen, advances can be made on the basis of understanding the relevance of human information processing components (for example, working memory, schemata, etc.) without specifying their underlying structural form (for example, as production systems, blackboard architectures, neural nets and so forth). I would go further and say that we may need to go beyond the currently accepted information processing components and allow for broad inclusion of all relevant social science components in this 'vertical slice' model of human performance.

In HCI, the most popular interactive behaviours to examine (if judged by published studies at least) used to be text-editing, a task so heavily studied and modelled that it has earned the somewhat derogatory title in some quarters of the 'white rat of human factors'.[2] It is only a matter of time before text-editing is replaced by web browsing (if that has not already happened) as the most studied interaction, but regardless of task, it is easy to see from such models that multilayered, architecture-independent analyses are rare. For example, one popular model of text-editing activity, judged more by citations than actual use in design, is based on the cognitive complexity theory (CCT) of Kieras and Polson (1985). This theory not only formally advocates the production system architecture of human cognition as a means of 'calculating' learning difficulties in transferring between text editors, it addresses only one level of activity for the system, that of correcting spelling mistakes in previously created text. The accuracy of the model is often held up as an example to other researchers and theorists in HCI, even though its utility to designers remains, some 15 years on, to be convincingly demonstrated, to put it mildly.

Predictive modelling techniques for HCI typically rely on identifying small units of behaviour, decomposing them into their assumed cognitive primitives, analysing them with respect to time and errors, and then developing an approximate model which accounts for performance within certain boundaries such as error-free expert performance. Such models of user behaviour with technology exist not only for text-editing but also in less extreme forms for menu navigation (Norman and Chen 1988), item selection with input devices (Card *et al.* 1978) and playing computer games (John and Vera 1992); so why not reading?

The crucial point is that reading could be equivalently modelled if ergonomists were to conceptualise it as narrowly as proofreading or item selection from a list of words as did Norman and Chen (1988). In

---

2   This descriptor was first seen by the author in an item on HICOM, the original electronic conferencing system of the British human factors community. The implication of the description is surely that text-editing tells us as much about HCI as a rat's performance tells us about being human. Depending on theoretical perspective that might mean a lot or a little, but given the absence of hard-line behaviourists in HCI research, one can only conclude that the writer intended it as a little.

equating reading with such activities, complexity is certainly reduced but range of application is surely curtailed. Accurate models of proof-reading might eventually lead to prescriptive principles for designing screen layout and manipulation facilities for such tasks (and only such tasks), rather like the GOMS model (Card *et al.* 1983) can theoretically aid the design of command languages for systems, but they are unlikely to prove extensible to many of the wider issues of electronic text design such as what makes a good or bad electronic text, or how a hypertext should be organised.

There is a school of thought that suggests that while such questions cannot be answered yet, the modelling approach is 'good science' and that sufficient progress in applied psychology will eventually be made by the accumulation and refinement of such low-level, predominantly mathematics-based models. Newell and Card (1985) argue that in all disciplines, hard science (that is, technical, mathematical) drives out the soft and that quantification always overpowers qualification. With reference to HCI they argue that psychology's proper role is to use its strengths and leave behaviour outside its remit to other disciplines. For Newell and Card, the domain of psychology covers human actions within the approximate timescale of 0.1 to 10 seconds. Anything smaller, they claim, is covered by the laws of physics, chemistry and biology, anything larger than this but less than a matter of days is covered by the principles of bounded rationality and the largest time frame of weeks to years is the proper subject matter of the social and organisational sciences.

Within the narrow timescale allowed psychology in their conceptualisation, Newell and Card propose that psychologists concentrate on 'symbolic processing', 'cycle times', 'mental mechanics' (which sound painful!) and 'short-term/long-term memory'. They accept that the bounded rationality time band covers many of the aspects of human behaviour relevant to HCI (and to humans in general it might be added) but 'their theoretical explanation is to be found in the interplay of the limited processing mechanisms of the psychological band and the user's intendedly rational endeavours' (p. 227).

Arguments that this form of psychology is too low level to be relevant to designers are partially correct, say the authors, but this problem will be overcome when a suitably all-embracing model of human information processing has been developed that can be applied wholesale to issues at the level of bounded rationality; that is, the level at which human activity is normally described and understood, for example, reading a book, writing a letter, driving a car and so forth.

This approach has been the subject of some strong criticisms (see for example, Carroll and Campbell 1986) and contradictory evidence. On the basis of interviewing designers of videotext systems, Buckley (1989) concluded that the type of model proposed by Card *et al.* (1983) was irrelevant to most designers. He claims that designers tend to avoid academic literature and specific experimental findings in favour of their own inter-nalised views of typical users and good design practice. Buckley states:

> The designers expressed concern about the ease of use of the dialogues and had some clear views of how system features under their control would affect users . . . But they did not report any use of traditional forms of human factors information which are expressions of the science base and are normally represented in research papers and design handbooks.
>
> (p. 183)

Instead, he found that designers in his research relied heavily on 'pre-existing internalised frameworks' (p. 184), which consist of primitive and weakly articulated models of users and their tasks that the system they were designing was intended to support. He goes on to emphasise the importance of providing information to designers in a form compatible with this style of working. Such findings are not unique, similarly doubtful views of the validity of formal models and standard human factors literature based on empirical findings have been expressed by other researchers who have interviewed or surveyed designers over the past decade (see for example, Dillon *et al.* 1993; Lund 1997).

Such findings partly confirm the conclusions drawn from the ADONIS study, where it was obvious that designers had some ideas of the users they were designing for, except that in this case, they were obviously also aware of some of the recommendations from the literature. This contrasts sharply with Buckley's designers, some of whom seemed surprised when he told them such a literature actually existed. Regardless of their familiarity with the literature, however, designers must have an idea of *who* they are designing for and *what* tasks the system will support; how else could they proceed? Their views are naturally partial and often intuitive. Therefore, making this conceptualisation more explicit and psychologically more valid in an appropriate way would seem to be of potential benefit to the software interface design world.

The second weakness in Newell and Card's argument is that it assumes the design world can afford to wait for an all-embracing cognitive model to emerge, while all around us, technological advancement accelerates. They counter this criticism with the somewhat surprising statement that technology does not advance as fast as many think it does.[3] Though they refer to paradigm interfaces, such as the standard graphical user interface

---

3   This is an observation that is worth taking seriously. Not only are more office applications built on desktop metaphors inherited from the 1980s, but also web design seems distinctly stuck in a repetitive groove.

found on many systems, as evidence that stability can be seen in design, it is less easy to see how their conceptualisation of HCI can either take us beyond current perspectives on usability or help ensure innovative applications are acceptably designed.

Regardless of the level of advancement, in the domain of reading at least, cognitive psychological models of the process exist which satisfy many of the criteria of hard, quantitative science but, as has been repeatedly pointed out, these just do not seem to afford much in the way of design guidance now. What seems to be required is a descriptive level above the information processing models advocated by Newell and Card but below the very high level descriptions of the bounded rationality approach favoured by information scientists. This is one of the levels we could expect to find in adopting an architecture-independent approach. In the case of reading and electronic text systems a suitably embracing framework would need to cover the range of issues from why an individual reads to how the screen can be best laid out, which would naturally induce inputs from a variety of research paradigms. However, these inputs would need to be organised and conceptually clarified in a manner suitable for designers.

Conclusions and the way forward

It has been argued in this chapter that many of the problems inherent in electronic text design spring from the lack of a design-relevant description of the reading process. Cognitive psychology and information science have been criticised for providing unsuitable levels of abstraction at which to describe the human behaviour relevant to design. This is, however, less a criticism of disciplines which do not exist to serve the needs of software designers, but more an indictment of human factors researchers' failure to provide their own theories. Barber (1988) remarks that ergonomics as a discipline has relied so heavily on theories borrowed from other disciplines that members of the human factors community see no need to develop their own and remain content to draw on standard psychological or physiological theories as required. If one could be guaranteed that standard theories could be drawn on this way and were never distorted by such application, this position might be tenable, but such guarantees cannot be provided by any theory.

Worse still, the tendency in the design world is for usability folks to accept a narrowly empirical method, aimed at gaining user inputs throughout the design process. While this is important, raw empiricism is a weak method, doomed to the constraints of time and financial budgets, even if one could imagine a form of empiricism devoid of theoretical insight (I cannot do so). Not only is theory unavoidable, it is the only hope we have for gaining greater foothold in the process of design where we cannot hope to succeed if we continue to be regarded as the people who are brought in at the end to determine usability.

The practical question then is what would a human factors practitioner or user experience specialist have added, for example, to the ADONIS design to make it more usable had she been involved as early as the specification stage? The simple truth of the matter is that deriving a more specific set of ergonomic criteria from the literature would have been difficult. The specification clearly included reference to most obvious variables. What would have helped is a user-centred design process of the form outlined in Chapter 2, involving stakeholder and user analysis, the setting of usability targets, iteration through prototypes, and continuous evaluations until the design targets were met.

The problem with this approach is that it is costly in terms of time and resources. What needs to be included is some means of constraining the number of iterations required. This is best achieved by ensuring that the first prototype is as close to the target as possible. Of necessity this would have involved carrying out user requirements capture and task analyses of readers interacting with journals and searching for articles in real scenarios. Output from such work would have been fed back to the designers to guide decisions about how the prototype interface should be built. Subsequent evaluation would then have refined this to an even better form.

It is almost certain that such work would have led to a better design than the current one, from which we can conclude that the type of knowledge generated by task analyses and prototype evaluations is directly relevant to design. The questions then become, what form of knowledge is this, at what level is it pitched, can it be extracted earlier and, more importantly, can a generalised form be derived to cover more than one specific reading situation?

An attempt to provide answers to questions of this type is made in the next three chapters. The primary means of providing them will be to examine the inputs made by the author to the design of real-world document systems as part of a variety of funded projects. By using this work as a background it is possible to identify the type of human factors inputs needed and found to be useful in real design projects.

If the results from any one document design project are to be generalisable to others, however, it is important to know how work on one text type differs from or is similar to others. Without such knowledge it would not be possible to make any meaningful generalisations about electronic text design from any one study or series of studies on a text. Unfortunately, there is as yet no agreed classification scheme for describing the similarities and differences between texts. To overcome this, a suitable classification scheme must be developed as a first stage of the work in deriving a framework for designing electronic texts. This is in line with other views. As de Beaugrande (1981) puts it: 'To adequately explore reading, a necessary first step is a firm definition of the notion of "text" – it is not just a series of sentences as one is often required to assume' (p. 297). He goes on to say: 'It follows that reading models will have to find control points in the reading process where text-type priorities can be inserted and respected' (p. 309).

To this end, the question of text type is addressed first and an investigation into readers' own classification systems of the world of texts is reported in the following chapter. This will be used to provide a basis to subsequent work and offer a means of generalising beyond the particulars of any one particular text type.

5 Classifying information into types: the context of use

There is no such thing as a moral or an immoral book. Books are well written or badly written.

Oscar Wilde, 1891

Introduction

Classification of concepts, objects or events is the hallmark of developed knowledge or scientific practice and, to a very real extent, typologies can be seen as a measure of agreement (and, by extension, progress) in a discipline. Fleishman and Quaintance (1984) point out that the major stumbling block in applying findings from the laboratory to the field within psychology is the lack of appropriate typologies (of tasks, acts, events and concepts). As I attempted to demonstrate in the literature review in Chapter 3, when it comes to studies of reading there is no standard reading text or task that can be used to investigate all important variables and thus, despite heavy reliance on proofreading short sections of prose (and there exists wide variability even in this type of task), we are not in a very good position to generalise effectively from experimental to applied settings.

de Beaugrande's (1981) call for an understanding of text types has been echoed often in the field of electronic text, where it is felt by some that a useful typology of texts could aid distinctions between potentially suitable and unsuitable electronic texts (McKnight *et al.* 1989). Such a typology would presumably provide a practical basis for distinguishing between the uses to which different texts are put and thus suggest the interface style required to support their hypertext instantiations.

There are a number of distinguishing criteria available to any would-be typologist of text, such as fiction and nonfiction, technical and nontechnical, serious and humorous, narrative and expository or, like Wilde, the well written or the badly written. These may discriminate between texts in a relatively unambiguous manner for some people and the list of such possible dichotomies is probably endless. Such discriminations, however, are not particularly informative in design terms. In this domain it is hoped (and even expected) that a typology could distinguish texts according to some usage criteria, leading to a situation where evidence can accumulate to the point where the designer could map interface features to information types reliably. In such a scenario, for example when designing any electronic document, it would be possible to select or reject certain design options that the typology suggests are appropriate or inappropriate for that text type. This contrasts sharply with current knowledge where any attempt to identify features known to have worked with, say, an experimental textbook might have no relevance to someone designing a web-based catalogue. Of course, as pointed out in Chapter 2, it is inappropriate to assume usability can ever be so prescriptively linked to interface variables in some form of inheritance mapping in a text or task scheme but the point here is that, ideally, a relevant typology should group information according to aspects of the reader–text interaction that support valid generalisation.

de Beaugrande (1980) defines a text type as 'a distinctive configuration of relational dominances obtaining between or among elements of the surface text, the textual world, stored knowledge patterns and a situation of occurrence' (p. 197).

This is a rather complex definition invoking the relationship between how the text looks, the readers' knowledge and experience as well as pointing to the context in which the text is met. The term 'textual world' is interesting and points to the subject matter of the document as well as its relationship to other documents. It is not clear to me, though, how relational dominances of this kind can be assessed. By way of illustration, de Beaugrande offers the following examples: descriptive, narrative, argumentative, literary, poetic, scientific, didactic and conversational. However, he freely admits that these categories are not mutually exclusive and are not distinguishable on any one dimension. As he says, 'many text types are vague, intuitive heuristics used by readers to tailor their processing to individual examples' (1981, p. 278).

This is interesting as it raises the possible distinction between text types as objective bases for discrimination according textual attributes and as subjective, process-oriented, reader-based perceptions. That is, can a typology be based on (more or less) objective criteria determinable by an examination of the published artifact or can one be drawn up only by taking account of readers' perceptions of type? And if the former, is it possible that constituent elements could be identified that supported the future classification of emerging digital forms? It would be convenient if the former was the case but it is more likely that, as in all things concerning humans, real insight is only gained by assuming the latter, that is, we will need an understanding of the readers' perceptions.

Typographers have a long-standing professional interest in text types, and, by and large, the discipline seems to categorise information on objective criteria, that is, typical typographic classifications relate to demonstrable attributes of the physical text. Waller (1987), for example, proposes analysing text types in terms of three kinds of underlying structure:

- topic structure, the typographic effects which display information about the author's argument, for example, headings;
- artifact structure, the features determined by the physical nature of the document, for example, page size;
- access structure, features that serve to make the document usable, for example, lists of contents.

According to this approach, any printed artifact manifests a particular combination of these three underlying structures to give rise to a genre of texts. This emergent fourth property is termed the conventional structure. As the name suggests these underlying structures give rise to the 'typical' text we understand as 'timetables', 'newspapers', 'novels' or 'academic journals'.

Like de Beaugrande, Waller describes the three structural components of his categorisation as heuristic concepts. However, for him this appears to have less to do with the reader approaching the artifact (although this is an element in Waller's categorisation) than with the means of providing a basis for genre description. From an ergonomic perspective it is therefore concerned less with the readers and their conceptualisation of the text than with layout, presentation and writing style. Thus a mapping of this analysis to the design of electronic texts is not logically determined even though the categories of structure outlined in this approach seem immediately more relevant to ergonomic interests than de Beaugrande's (and the concept of access structure, which evokes usability, has certainly been embraced in discussions of electronic text design (for example, Duffy *et al.* 1992).

Borrowing from the work of Seymour Chatman (1990), Smoliar and Baker (1997) analysed three distinct text types in a hypermedia context: descriptions, arguments and narratives. They argue that this high-level categorisation provides the kind of organisational framework that can facilitate production or writing of hypermedia documents as well as aiding readers to transform and make sense of a writer's text. As such, this could be the basis of a form of writing/reading aid for more usable document design. It is not easy to see how these three forms alone would prove sufficient to enable all these authors imagine, but their argument emphasises further the importance of conceiving the possibilities for electronic text outside the artificial boundaries of assumed normative forms.

Interdisciplinary boundaries disappear very quickly when one starts to think seriously about text types. Cognitive psychologists have taken increasing interest in the relationship between so-called 'typographical' features and the reading process (for example, Hartley 1985; Wright 1998) and typographers look to psychology for theoretical explanations of typographic effects on humans. This interdisciplinarity has not always been so positive, however. Waller (1987) reports on a long-standing debate in the typographic discipline on the value of applied psychological research (largely the univariate, experimental model manipulating text attributes such as line-length or font). Criticisms of this type of work, typified early on by people such as Tinker (1963), have existed since the 1930s (for example, Buckingham 1931), and in nature and content largely resemble current arguments in the field of HCI where the value of human factors experimental research to system designers is often questioned. One can only hope such rigid mindsets are a thing of the past, but it may not be that simple.

Ergonomics invariably attempts to draw heavily on psychological theories in extending its reference beyond the immediate application and it is clear that some work on text groups exists in contemporary psychology. In psycholinguistics, for example, van Dijk and Kintsch (1983) use the term 'discourse types' to describe the superstructural regularities present in real-world texts such as crime stories or psychological research reports. Their theory of discourse comprehension suggests that such 'types' facilitate readers' predictions about the likely episodes or events in a text and thus support accurate macroproposition formations. In other words the reader can utilise this awareness of the text's typical form or contents to aid comprehension of the material. In their view, such types are the literary equivalent of scripts or frames and play an important role in their model of discourse comprehension. While they do not themselves provide a classi-fication or typology, this work is directly relevant to the design of electronic texts, as is discussed in detail in a later chapter.

Wright (1980) described texts in terms of their application domains and grouped typical documents into three categories: domestic (for example, instructions for using appliances), functional (for example, work-related manuals) and advanced literacy (for example, magazines or novels). She used these categories to emphasise the range of texts that exist and to highlight the fact that reading research ought to become aware of this tremendous diversity. This is an important point, and one central to ergonomic practice. Research into the presentation and reading of one text may have little or no relevance to, and may even require separate theoretical and methodological standpoints from, other texts. Consider, for example, the range of issues to be addressed in designing an electronic telephone directory for switch-board operators with no discretion in their working practices, and an electronic storybook or 'courseware' for children between the ages of 8 and 10. Rigid implications on interface design drawn from one will not easily transfer to the other unless one at least possesses the appropriate perspective for understanding the important similarities and differences between them. A valid typology of texts should be useful in such a situation as researchers from a range of disciplines have suggested.

Reader-perceived distinctions between texts

What all the classifications outlined above failed to take account of, however, is the readers' views of these issues and it is with this concern in mind that a colleague and I decided to carry out an investigation of readers' perceptions of text types (Dillon and McKnight 1990). This marked a first attempt at understanding types purely according to reader-perceived differences rather than theory- or theorist-derived distinctions. In so doing it was hoped that any emerging classification criteria would provide clues as to how electronic documents can best be designed to suit readers, an intention not directly attributable to any of the aforementioned categorisations.

It was not immediately obvious how any researcher interested in text classification schemes should proceed on this matter, however. Few categorisers to date have made explicit the manner in which they derived their classifications. It seems that regardless of theoretical background, most, if not all, have based their schemes on their own interpretations of the range of texts in existence. True, their classifications often seem plausible and the knowledge and expertise of all the above-cited proposers is extensive; yet it is impossible to justify such an approach in the context of system design. It could be argued that readers do perceive the manifestation of the different structures or heuristics even if they cannot reliably articulate them. After all, the conventional genres are real-world artifacts. However, this can only be assumed, not proven as yet. This section reviews the Dillon and McKnight work and in so doing reports some readers' perceptions of text types. An extension of this work exploring the more general concept of information space is reported by McKnight (2000).

In the first instance it must be said that we soon realised that recognising the need for a more objective means of classification was one thing, identifying a suitable technique to enable this was another. To derive a reader-relevant classification necessarily requires some means of measuring or scoring readers' views. The available methodological options are techniques such as interviewing, questionnaires or developing some form of sorting task for readers to perform on a selection of texts. The case for using a questionnaire is flawed by the absence of any psychometrically valid questionnaire on this subject and the effort involved in developing one within the timescales of any one investigation.[1]

Interviewing individuals is a sure method of gaining large amounts of data. However, making sense of the data can prove difficult both in terms of extracting sense and in overcoming subjective bias on the part of the interviewer. While the latter problem can be lessened by using two skilled interviewers rather than one, and careful design of the interview schedule, the former problem is more difficult to guard against in this context. Text classification, as a subject, could be sufficiently abstract as to cause interviewees problems in clearly articulating their ideas in a manner that would

---

1 A questionnaire developed according to sound psychometric principles is a lengthy task involving stages of item generation, selection, piloting and analysis, possibly through several iterations before a reliable and valid tool is developed (Oppenheim 1966). This distinguishes questionnaires from the more loosely created 'questions on a page'-type surveys common to human factors or market research.

support useful interpretation of the data. With these issues in mind we decided to consider some form of sorting task.

Repertory grid analysis was eventually chosen as the most suitable technique of this type for eliciting suitable data. Developed by George Kelly (1955) as a way of identifying how individuals construe elements of their social world, personal construct theory (PCT) assumes that humans are basically 'scientists' who cognitively (or more accurately, in Kelly's terms, 'mentally') represent the world and formulate and test hypotheses about the nature of reality.

The repertory grid technique has been used for a variety of clinical and nonclinical applications (for example, studying neurotics: Ryle 1976; job analysis: Hassard 1988) and has been applied to the domain of HCI, particularly with respect to elicitation of knowledge in the development of expert systems (Shaw and Gaines 1987). Its use as an analytic tool does not require acceptance of the model of man which Kelly proposed (Slater 1976). However, the terms Kelly used have become standard. Therefore, one may describe the technique as consisting of *elements* (a set of 'observations' from a universe of discourse), which are rated according to certain criteria termed *constructs*. The elements and/or the constructs may be elicited from the subject or provided by the experimenter depending on the purpose of the investigation. Regardless of the method, the basic output is a grid in the form of *n* rows and *m* columns, which record a subject's ratings, usually on a five- or seven-point scale, of *m* elements in terms of *n* constructs.

The typical elicitation procedure involves presenting a subject with a subset of elements and asking her to generate a construct which would meaningfully (for her) facilitate comparison and discrimination between these elements. The aim is to elicit a bipolar dimension which the subject utilises to comprehend the elements. In the present context, all participants were given a similar set of texts (the elements) which they used to generate distinguishing criteria (the constructs). As constructs are elicited and all elements subsequently rated on these, a picture of the subject's views and interpretations of the world of documents emerged. What follows is a summary of the results published by Dillon and McKnight (1990).

Readers, texts and tasks

Six professional researchers had grids elicited individually. Elements, which were identical for all, consisted of nine texts:

a newspaper (*The Independent*, a UK 'quality' daily broadsheet)
a manual (*MacWrite Users Guide*)
a textbook (*Designing the User Interface* by Ben Shneiderman)
a novel (*Steppenwolf* by Herman Hesse)
a journal (*Behaviour and Information Technology*)
a catalogue (*Argos Catalogue Spring* – personal and household goods)

a conference proceedings (*ACM SIGCHI Conference*)
a magazine (*The Observer Sunday Colour Supplement*)
a report (A research institute technical memo)

Constructs were elicited using the minimal context form (Bannister and Mair 1968), which involves presenting people with three elements, known as the triad, and asking them to think of a way in which two of these are similar and thereby different from the third. When a meaningful construct was generated the two poles were written on cards and placed either side of a 1–5 rating scale on a desk. Researchers then rated all the texts according to the construct, physically placing the texts in piles or individually on this scale.

Results

Full details of the analysis can be found in Dillon and McKnight (1990). It is sufficient here to describe the resulting grids from the FOCUS program used (Shaw 1980). A focused grid for one subject is presented in Figure 5.1. The grid consists of the raw ratings made by the subjects with the element list above and the construct list below. The FOCUS program automatically

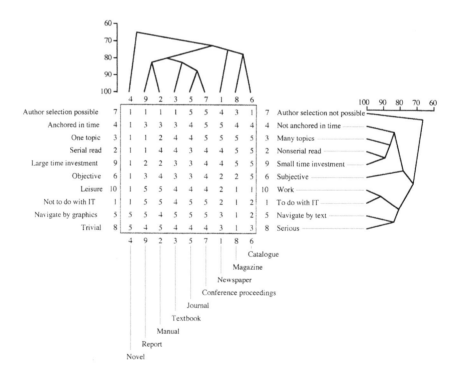

Figure 5.1 FOCUSed grid for one subject.

reorders these to give the minimum total distance between contiguous element and construct rating columns. Dendrograms are constructed by joining elements and constructs at their appropriate matching levels.

In Figure 5.1, the elements are on top and the constructs are to the right of the reordered ratings. The matching levels for both are shown on adjacent scales. High matches indicate that the relevant elements share identical or similar ratings on the majority of constructs or the relevant constructs discriminate identically or similarly between the majority of elements. Thus in Figure 5.1 it can be observed that the journal and conference proceedings match highly and that the novel is least similar to the others. Criteria such as 'Work – Leisure' and 'To do with IT – Not to do with IT' offer the highest match among the constructs elicited, while 'Single author selection possible – not possible' is the lowest match. In other words the journal and the conference proceedings are seen as very similar to each other but very different from the novel by this reader, and every time a text is described as work related it also tends to be described as being about information technology. By proceeding in this manner it becomes possible to build up a detailed picture of how an individual construes texts. In the present study all six grids were analysed (focused) together as one large grid. The focused collective grid is presented separately in the element and construct trees in Figures 5.2 and 5.3.

The texts

The texts (elements) clustered into three distinct groups (see Figure 5.2). These were the work-related, the 'news'-type texts and the novel. The highest match was between the conference proceedings and the journal (90.2 per cent), followed by the newspaper and the magazine (85.1 per cent). Basically

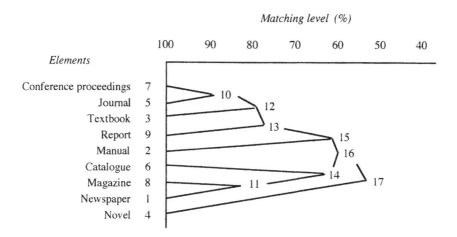

Figure 5.2 Dendrogram of element clusters for all readers.

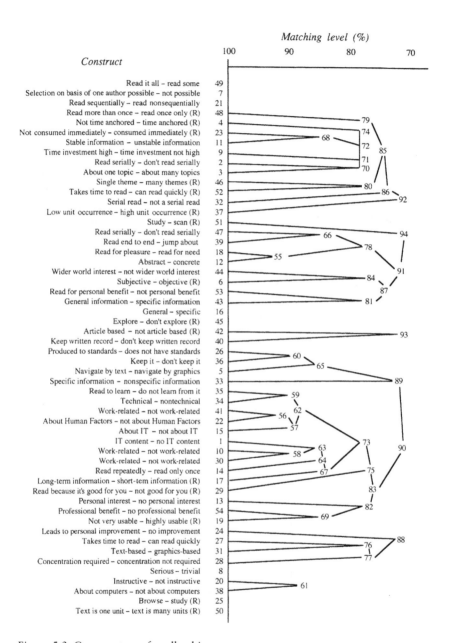

Figure 5.3 Constructs set for all subjects.

this means that any time, for example, the journal was rated as being high or low on construct X then the conference proceedings were rated similarly. The textbook and report both joined the first cluster at more than the 82 per cent matching level. This cluster eventually incorporated the software manual at the 62 per cent level, suggesting that while this manual shared some of the ratings of the other elements in that cluster it was noticeably different from them on certain constructs.

The catalogue matched the newspaper and magazine at 69.4 per cent suggesting that it is perceived as similar in many ways to those types of text. The novel, however, was the last element to be incorporated in a cluster, only linking with other elements at the 53.2 per cent level, by which time all the other elements had formed one large cluster. This suggests that it was perceived as a unique text type among all these elements.

The readers' distinguishing criteria

Fifty-four constructs were elicited from this sample. In order to ensure that only tight clusters were identified, a minimum matching level of 70 per cent was defined as the criterion. This is somewhat arbitrary and at the discretion of the investigators, the lower the criterion the more matches are produced until eventually all constructs would match at some point. The construct dendrogram is presented in Figure 5.3. Dillon and McKnight (1990) identified three major construct clusters from this analysis which are outlined below.

Cluster 1: Work-related material   This cluster described texts which are work-related, about human factors or information technology, contain technical or specific information and would be read for learning or professional purposes.

Every researcher distinguished between work-related and personal reading material. All of their constructs about this distinction matched up at the 77.7 per cent level. Obviously, construing texts in terms of their subject matter and relevance to work was common to all these researchers. The constructs 'reading to learn' and 'technical' matched at 88.8 per cent and joined up with 'work-related'/'about human factors'/'IT-related' at that level too. Also contained in this cluster were 'read repeatedly' and 'long-term information', matching a work-related subcluster at the 83.3 per cent level. An element that was prototypical of this construct cluster was the journal. A very poor match with this cluster was observed for the newspaper and the magazine.

Cluster 2: Personal reading material   This contained texts that were seen as personal reading material, containing general or abstract information that would be read in a serial fashion.

The highest match in this cluster was between the constructs 'abstract – applied' and 'reading for pleasure – reading for need' which matched at the 94.4 per cent level. The next highest was at the 83.3 per cent level between 'serial' and 'read from end to end'. These pairs of constructs then joined at the 77.7 per cent level. The constructs 'personal benefit', 'general information', 'subjective', and 'wider-world interest' all matched at the 77.7 per cent level. These subclusters all joined at the 72.2 per cent level.

These constructs suggest that certain texts are seen as more personal than work-related and contain information that is general in nature or subjectively interesting. The presence of constructs that indicate they are read in a serial fashion would indicate texts that are not intended for reference but for complete reading. A text that closely matched most of these descriptors was the novel. A very poor match with these constructs was the catalogue.

Cluster 3: Detailed lengthy reading material   This cluster described texts that were seen as having one main subject or topic, the content of which is stable and requires a high time-investment to read. Such texts are also characterised by serial reading.

The highest match in this cluster was between the constructs 'not immediately consumable' and 'contains stable information' which matched at the 83.3 per cent level. The constructs 'read serially' and 'one topic' matched at the 77.7 per cent level, as did 'time investment to read is high' and 'single theme'. The constructs 'read more than once' and 'not anchored in time' matched at 77.7 per cent and all these constructs were joined by another construct 'serial' at this level too. The final construct in this cluster above the criterion level was 'low occurrence of separate units in the text' which joined all of the other constructs at the 72.2 per cent level. A text that closely matched these constructs was the novel; the newspaper and magazine were typically rated as the opposite of these.

The results demonstrate that people's manner of construing texts is complex and influenced by numerous factors. Clear distinctions between texts such as 'fiction and nonfiction' are too simplistic and superficial. On a psychological level individuals are more likely to make distinctions in terms of the type of reading strategy that they employ with a text, its relevance to their work or the amount/type of information that a text contains. This is not a basis for classification used by many proponents of text typologies.

While the terms or descriptors employed and their similarities or differences (the face validity of the output) are interesting, it is their treatment of the elements that is ultimately important. The elements – textbook, journal, conference proceedings and report – were all matched highly, forming a particular cluster of text types. The magazine and newspaper were also matched highly. These are reasonable groupings between what may broadly be termed 'work' and 'leisure' texts. The novel is the one text type that matches least well with all the others and, once again, this appears sensible. Examining the constructs that distinguish between these texts can shed more light on the classification criteria employed by these readers.

The journal and textbook types were described, unsurprisingly, as work-related, about human factors or IT, containing specific or technical information, and were read for professional benefit or in order to extract specific information. They were likely to be read more than once and be of long-term rather than immediate use or relevance. This distinguishes them from the other two element clusters which were more likely to be described as read for leisure and containing general, subjective or non-technical information. The novel was further distinguished from the newspaper and magazine by the need to invest a lot of time and read it serially and completely.

On the basis of these results we argued that for any given context of use texts seem distinguishable on three levels which any one construct may reflect:

- Why they are read, for example, for professional or personal reasons, to learn or not, out of interest or out of need, etc.
- What type of information they contain, for example, technical or nontechnical, about human factors or not, general or specific, textual or graphical, etc.
- How they are read, for example, serially or nonserially, once or repeatedly, browsed or studied in depth, etc.

By viewing text types according to the various attributes of these three levels of use it is easy to distinguish between, for example, a novel and a journal. The former is likely to be used for leisure (Why), contain general or nontechnical information (What) and be read serially (How), whereas the latter is more likely to be used for professional reasons (Why), contain technical information which includes graphics (What) and be studied or read more than once (How).

Importantly, there is an individualistic aspect here. The same text may be classified differently by any two readers. Thus a literary critic is likely to classify novels differently from a casual reader. Both might share similar views of *how* it is to be read (for example, serially or in depth) but differ in their perceptions of *why* it is read or *what* information it contains. The critic will see the novel as related to work while the casual reader is more likely to classify it as a leisure text. What it contains will differ according to the analytic skill of the reader, with a critic viewing, for example, Joyce's *Ulysses* as an attempt to undermine contemporary English and the casual reader (if such exists) seeing it as a powerful stream-of-consciousness modern work. Neither is wrong; any classification of texts based on psychological criteria must, by definition, allow for such individual differences in perception and meaning.

Readers may vary their classification of texts according to tasks. Several researchers remarked that some texts could possibly be classed as work-related and personal reading depending on the situation. An obvious example of this occurs when someone reads an academic article that is both relevant to one's work and intrinsically interesting in its own right. For individuals whose professional and personal interests overlap such an occurrence was common. The present categorisation of texts allows for this by placing emphasis on the motivation for reading (the Why axis).

Allowing for both between- and within-subject variance reflects the underlying psychological nature of the categorisation. A classification based solely on demonstrable objective distinctions is likely to have either a very limited sphere of application outside of which it loses relevance or be as simplistic as distinctions of the form paperback–hardback or fiction–nonfiction, etc. The present classification does not therefore provide a formal typology of texts but it does offer a way of distinguishing between them and analysing readers' perceptions. It suggests that three aspects of the text are important to the reader in categorising a document or published artifact.

From text types to context types: implications for design

How does all this relate to the development of electronic text systems? Theoretically, at least, one could seek to determine the complete membership of each attribute set by identifying all possible *hows*, *whys* and *whats* through some form of scenario generation or task analysis and subsequently plot the range of texts that match any combination of the three. However, such a level of analysis may be too fine-grained to be worth pursuing. The perspective proposed here is best seen as a simple representation of the factors influencing readers' perceptions of texts. Characterising text according to *how*, *why* and *what* variables can provide a means of understanding the manner in which a given readership is likely to respond to a text. So, for example, it could be used for understanding the similarities and differences between a telephone directory:

> Why: to contact a friend;
> What: specific numeric data;
> How: skim and locate;

and a novel:

> Why: leisure;
> What: large text, no graphics;
> How: serial, detailed read;

and thus quickly facilitate sensible decisions about how they should be presented electronically. Given what we know about reading from screens and HCI, the position of a novel in such a classification system would suggest that an electronic version would not be used frequently whereas an electronic telephone directory may usefully be designed to ease the process of number location or aid searching with partial information.

At a gross level, such a classification may serve to guide decisions about the feasibility of developing a usable electronic version of a text type. Where the likely readership is known this can act as a stimulus to meaningful task analysis to identify how best to design such texts. In the example of the novel above, this might reveal that the novel is required for teaching purposes, where sections need to be retrieved quickly and compared linguistically with other sections or writers. Obviously this would alter the *how* and *why* attributes of the previous classification, indicating that an electronic version is now more desirable, for example,

> Why: education in literary studies/text analysis
> What: large text, no graphics
> How: location, short section reading and comparison.

This highlights how such a categorisation could lead to a more informed analysis of the design potential than deceptive first impressions and common sense suggest. Hypertext novels could in fact be a boon to teachers in a variety of educational scenarios, yet even expert human factors professionals might not see that at first glance. Norman (1988), for example, when discussing the potential for hypertext applications in general said: '[Hypertext] is a really exciting concept. But I don't believe it can work for most material. For an encyclopaedia, yes: or a dictionary; or an instruction manual. But not for a text or a novel' (p. 235).

Regardless of what he means by 'a text' in this quotation, it is clear that a more critical examination of the *why* question, the motivation or reason for reading (in this case novels) indicates that a hypertext version could work. Obviously even human factors experts cannot be relied on to see all the user implications of any technology in advance!

By knowing more about why individuals access texts, how they use them in terms of reading strategies and the distinctions they make between the information type presented, it should prove possible to be more specific about the question of text types. Traditionally, the *how* question has been the domain of the psychologist, the *what* question the domain of the typographer and the *why* question has been largely ignored. The results from this study strongly suggest that this is a mistake.

Texts possess more than purely physical properties. Readers imbue them with values and attribute them roles in the support of a host of real-life acts in innumerable work and leisure domains. This is not immediately obvious from a physical examination of a text and indeed cannot be understood by any reader immediately. Rather, these are acquired skills, perceptions shaped by years of reading experiences and the responses of authors and publishers in attempting to provide support for reading.

Reading is purposeful and intentional. Texts can support many such interpretations for the reader depending on context. As I have attempted to show in the present chapter, what is a relaxing serial read for one reader may be a jumping quick scan for another. As such, it is nonsense to assume we can devise prescriptive rules for designing standardised electronic versions of certain text types such as novels, journals or catalogues (or other conventional genres). One should perhaps realise that we cannot typify texts for readers, only generalise reading contexts with their constituent interactive components of reader, task, text and environment. Thus the genres we require for electronic texts are ones embodying criteria of usage such that they are grouped according to the tasks they support, the reasons for the interaction and the nature of their contents. In other words there is a coupling between a text and the reader and this may shift as the reader's task alters. Reliance solely on existing conventional genres is misleading since any one genre may exist in multiple contexts, where it is repurposed as needed.

It is conceivable that utilising this perspective on information types could support the design of new text types. In other words, by conceptualising the reader–text context of use appropriately according to *why*, *what* and *how* criteria, documents could be created that support the reader's interaction with the information. This would be a natural extension of user-centred design methods – creating documents specific to user needs rather than seeking always to modify existing types. However, this is unlikely to be so simple since, as I will attempt to show later, existing genres are potent formats that in themselves impose and support much by way of use. In a very real sense, therefore, what is being proposed is not so much a classic text typology than a means of relating texts to contexts.

Conclusions and the way forward

The variance in texts that readers regularly utilise was identified as a natural starting point for any analysis of the real-world reading process, and an area requiring specific attention in the context of electronic text systems. The present chapter reviewed several approaches to describing texts in a manner suitable for discussing the design of electronic versions. To this end, a description based on three reader-perceived characteristics seems most appropriate: the *why*, *what* and *how* aspects.

These aspects represent readers' own classification criteria and also offer a means of describing texts in a way that is directly related to designing electronic versions; that is, if a book is accurately described according to these criteria it should lead to specific issues for consideration in design in a way that would not necessarily be the case for a description based on more traditional criteria. Specifically, by focusing attention on how a text is read immediately leads to a consideration of task issues; by examining what a text contains the questions of content and structure are addressed; and by focusing on the why aspect the context and motivation of reading are highlighted. Furthermore, by appreciating the differences in texts in these terms one is in a better position to judge the likely relevance of experimental findings on one text to another. The real test of this approach, however, is the extent to which meaningful data can be derived from such a classification. In the following section, empirical applications of this approach are described.

6 Capturing process data on reading: just what are readers doing?

> From the moment I picked up your book until I laid it down, I was convulsed with laughter. Some day I intend reading it.
>
> Groucho Marx, 1945

Introduction

The reader-elicited text descriptors outlined in the last chapter provide a starting point for a consideration of the design issues for electronic text. They reflect both cognitive and behavioural aspects of the reader–text interaction rather than more common classifications based on publication genre or subject matter. Focusing attention on why and how texts are read, as well as on the reader's views of what texts contain, immediately brings forth issues related to task, motivation for reading, readers' models of the information space and the reading situation or context of use; factors certain to be of importance in the ultimate success or failure of any presentation medium.

It is a simple enough matter to describe a text according to these three criteria if the description consists only of the type of one-liner provided in the examples of the last chapter. However, such descriptions are not enough to provide a basis for specifying software where details of a more precise nature are required (see for example, Sommerville 1997). Furthermore, merely describing texts in this way as a result of introspection or best guess on the part of the designer or human factors expert is far from optimum (though such uses might be appropriate for the initial consideration of issues in a design prior to formal specification or user evaluation). It is clear that designers are normally not the best source of information on how typical users of their products view the situation. What is required, therefore, is the demonstration that this approach can be utilised to gather evidence of reader behaviour (particularly process data) and that the resulting output has relevance to system design.

The range of texts analysed

In the course of design work on electronic documents, my colleagues, doctoral students and I have analysed multiple document types: academic journals, software manuals, research project material (a collective description of the reports, data and documented communications of a team of researchers on one project), a handbook of procedures for a multinational manufacturing company, digital newspapers, websites and online image databases. You can still find detailed accounts of some of this work elsewhere, for example, the research project material is covered in McKnight *et al.* (1992) and a full account of my student's work on digital newspapers can be found in Vaughan (1999).

Much of the work reported in this book was carried out in parallel with, and often as part of, a series of research and development projects designing and evaluating electronic documents at HUSAT in Loughborough, University, at Indiana University and now at the University of Texas at Austin. The investigation into journal usage described here had direct relevance to real design issues for the team designing an electronic journal application and, in fact, the investigation was driven by the need for information of this kind. Software manuals were selected for analysis as it was felt that they are a frequently used text for which electronic versions partly exist (in the form of online help facilities). Furthermore, they satisfy the criterion of distinction from journals as a text type according to the results of the repertory grid study, thereby offering a useful indication of the breadth of application for the *why*, *what* and *how* approach to simulated usage.

Describing text usage: the situated simulation method

As mentioned in Chapter 3, process data of reading are hard to obtain reliably and in an unobtrusive manner. The standard approach of most social scientists is to devise an experiment to answer any question. In the case of reading studies this has led to a bias in favour of outcome measures such as speed or accuracy of reading performance, or comprehension level (although even the measurement of that particular outcome is the source of some debate). Unfortunately, where the classical experimental method (by which is meant the isolation, control and manipulation of specified variables) proves difficult or impossible, it is often the problem that is considered ill-specified rather than any shortcomings of the formal experimental method that are exposed. This can be appreciated further by examining some of the more intrusive methods that reading experimentalists have employed to capture process data in a supposedly justifiable methodological manner. The use of bite bars and forehead rests (see for example, McConkie *et al.* 1985) to hold readers' heads steady while text is passed before their eyes might sound sublime or ridiculous, depending on one's predilection, but there may be some reasonable doubts about their appropriateness for ecologically valid work.[1]

To insist on a laboratory-derived experimental approach to capturing process data here would not only require eye monitoring of the kind so beloved by legions of experimental psychologists, but also demand a means of recording all physical manipulations of the text, possibly involving several video cameras, and an assurance from the readers that they would not move too fast or into a position that blocked out the camera's view. This is hardly a situation likely to induce relaxed examination of the reading process with interested readers and would therefore leave the current problem of how texts are read largely intractable, thus lending support to Wittgenstein's (1953) argument that in psychology: 'the existence of the experimental method makes us think that we have the means of solving the problems that trouble us; though problem and method pass one another by' (p. 232).

The issues of why, what and how texts are read are our problems here, but they do not in themselves determine the data capturing methods to be employed in their solution. In seeking answers to such questions I soon realised that the classic experimental method was not going to serve us well here, although no single alternative technique offered the means to answer the questions being posed. However, the questions were legitimate; therefore, it was decided that a mixture of investigative procedures must be employed.

Interviewing and observing a selection of relevant readers seems the most suitable means of gathering relevant data; that is, why people read certain texts and what they typically expect the documents to contain? As mentioned previously, the advantages of interviewing are that it facilitates the elicitation of data that are difficult to obtain from more formal methods, as well as supporting the opportunistic pursuit of interesting issues. The problems and limitations of interviewing as a data elicitation method, however, have been well documented (for example, Kerlinger 1986). Common problems include deriving a suitable structured schedule that ensures comparability of results without constraining the interviewer, or scoring/coding what can be 'messy' data in a reliable manner.

In the work described here, the criteria derived from the repertory grid study outlined in the previous chapter provided a form of structure for eliciting and analysing the data. Thus a core set of issues to cover with every reader is identified and a standard means of categorising answers can be

---

1 The term 'ecologically valid' describes work that seeks to reflect the real-world aspects of a task rather than work that takes the issue under investigation out of its natural situation of occurrence and places it in isolation. It is a term that causes a lot of debate among human scientists, not least in the present context. A recent article of some colleagues of mine was criticised for using the expression to describe a measure we had taken. The referee never explained their objection but merely wrote 'Ecological validity? Not in this journal please'!

obtained. The interview questions are typically devised by discussion, in the first instance among the team of designers involved in building a journal database. This ensured that no important issues were overlooked and that the data were of direct use to the design process. However, this also meant that for each investigation, the exact questions are never the same since they result from negotiations specific to the time and context of the investigation. This might be seen as a source of weakness and in objective terms it is surely not 'good science.' In defense of the approach, though, the core questions (why, what and how) and their immediate derivatives are always addressed in a similar manner (as will be shown) and it is clear that the flexibility to add or alter some aspects of the investigation to suit the design situation could be seen more as a strength than a weakness of this approach, particularly in the unstable context of a software design process.

Interviewing alone, however, will not sufficiently answer the question of how a text is read. Certain issues or topics are not easy to describe adequately using only verbal means. To obtain suitable information it was decided that simulated usage or task performance with concurrent verbal protocols would suitably complement the structured interview approach. The basic idea here is to ask readers to simulate their typical interaction with a text from the moment of first picking it up to the time of finishing with it, articulating what they are attending to with respect to the text at all times.

This method simulates probable task behaviour from the reader without resorting to elaborate means of investigation or generating masses of low-level data of the type certain to have emerged if readers had been set formal tasks and their interactions recorded on video tape. Readers in these investigations are asked to look at a selection of (predetermined) relevant texts and to examine each one as they normally would, for example, if browsing them in the library, their office, or sitting at home in their favourite armchair. They are prompted to articulate what information they cue into when they pick up a text, how they decide what is really worth reading and how they read, say, journal articles that are selected for individual use. They repeat this simulation for several texts of that genre until a consistent pattern emerges. This determination is really the investigator's call, but what follows is crucial. At this point, the reader is asked to confirm the accuracy of the investigator's interpretation of their reading style. If both agree, the process is complete. If there is disagreement, the process continues until agreement is reached. In this way, the reader becomes part of the investigation procedure. What constitutes the description of the reading process is effectively negotiated between investigator and reader, and the reader can, at all stages, correct the interpretations of the observer. The process concludes only when agreement is reached.

The main data source in this method is the verbal protocol. Like interviewing, much has been written about the use of verbal protocols in psychological investigations. The main issue of contention is the extent to which they can be said to reflect reliably the speaker's underlying cognitive processes or are merely a reflection of what the verbaliser thinks is appropriate and/or what they think the experimenter wants to hear (see for example, Nisbett and Wilson 1977).

Ericsson and Simon (1984) have developed a framework for the use of verbal protocols and related it to psychological theories of articulation which suggest that for tasks where subjects are required to describe what they are doing or attending to in real time (concurrent verbal reporting), objections on the grounds of inaccuracy or unreliability of self-reports rarely apply. Problems of accuracy are more likely to occur during retrospective verbal reporting ('This is how I did it . . . '), as human memory is fallible and subject to post-task rationalisation (a concern that should give us pause when considering the use of post-task protocols in usability studies); or when reporting on how their own mental activities occurred ('I had an image of the object and then recalled . . . '). In other words, when humans report what they are doing or trying to do during the performance (or simulation) of a task, and are not requested to interpret their own thinking (as in introspection), there are no *a priori* grounds for doubting the validity of their comments. Obviously subjects may lie or deliberately mislead the experimenter but this is a potential problem for all investigative methods requiring the subject to respond in a nonautomatic fashion. The point here, however, is that people can reliably report what they are thinking (that is, current contents of working memory) but are less likely to be able to do so for how they came to be thinking of it (that is, what cognitive processes brought these contents to working memory).

Verbal protocol data of this form are regularly elicited in HCI studies and have been used to good effect in analysing the influence of various interface variables on users' perceptions of, and performance with, a system. Concurrent verbal protocols are used with this method, not to provide deep insight into the cognitive process of the reader but to provide a verbal accompaniment to the behavioural act of text manipulation and usage. What is different from more typical verbal protocol analyses is the role of the verbaliser in clarifying their aims and behaviours.

To control for bias or potential limitations in the ability of one investigator to capture all relevant data, another one was employed for the journal study (the first use of this approach), thus providing a corater for the elicited data and maximising data capture. All interpretations and conclusions were checked with this observer also before final agreement was reached. After the first investigation, because of the high level of agreement between investigators and the lack of difficulty noted in one investigator recording the data with appropriate readers, this was felt to be unnecessary for further work. However, it would be wise for first-timers to work in pairs until familiarity with this form of data capture is gained.

The situated simulation method in action

In both cases the investigation was similar. Fifteen readers participated – this was originally deemed sufficient from the first study when it started to become clear that no new insights were being gained as more readers were being interviewed. It is possible that most major issues could be captured with less than this number but, alternatively, it is also likely that a fully accurate picture would emerge with many more participant readers.

Readers were interviewed individually and presented with a range of texts of the type under investigation. The basic procedure involved an interview to collect information on why readers used journals and what types of information they thought such a text contained. Typical prompts at this stage involved variations of the *Why* and *What* aspects, pursuing points as they developed and concentrating on those areas that have any impact on usage.

Readers then interacted with a sample of relevant texts, simulating their typical usage according to their expressed reasons, articulating what they were attending to as they did so. This provided the *How* data. Readers were prompted as necessary by the interviewer. After describing and simulating their typical usage of the texts, the interviewer described his impression of their style and sought feedback from the reader that it concurred with the readers' own views. When agreement had been reached, that is, the reader and interviewer both agreed that the representation of usage was accurate and adequate, the investigation ended.

The purpose of double-checking one's interpretations with the participants is crucial. What you are seeking here is elusive and it is easy to miss or misinterpret rapid actions and comments. By sharing your interpretations with the reader you can confirm that you are on the right track. Such a process is central to gaining a true user-centred perspective and is a standard part of what is generally referred to now as a process of contextual inquiry (Beyer and Holtzblatt 1998).

Describing usage according to Why, What and How attributes

To facilitate comparison between the two texts and to show how the general situated simulation approach operates, each attribute of usage is presented in turn and the data from both texts are combined.

Why read a text?

The most frequently stated reasons for accessing academic journals are summarised in Table 6.1. while the equivalent data for software manuals are presented in Table 6.2, and the data are rather self-explanatory. Almost

Table 6.1 Stated reasons for using journals

Why use a journal?	No. of subjects	(%)
Background material for work purposes	11	(73)
Updating one's knowledge	7	(46)
Personal interest	3	(20)
On recommendation	2	(13)
Following up references	2	(13)

Table 6.2 Stated reasons for using software manuals (from Dillon 1991)

Why use a manual?	No. of subjects	(%)
For reference/How do I do this?	11	(73)
How to get started	10	(67)
When in trouble	8	(50)
For a summary of package's facilities	5	(33)
Aid for exploring software	2	(13)
As a guide before buying	1	(7)
For detailed technical information	1	(7)

all readers distinguished between problem-driven journal usage, where work demands require literature reviews or rapid familiarisation with a new area, and personal usage, where journals are browsed in order to keep up with latest developments in one's area of expertise or interest, the former being cited more frequently than the latter.

In more specific terms, journals are accessed for: personal interest, to answer a particular question (for example, 'what statistics did the authors use and why?'), to keep up with developments in an area, to read an author's work and to gain advice on a research problem. In other words, there are numerous varied reasons for accessing material in journals apart from just wanting to 'study the literature'.

In terms of software manuals, readers stated numerous reasons for using manuals though there was a large degree of consistency in their responses. The three main reasons (reference, introduction and when in trouble) were all offered by at least half the readers. These highlight the problem-driven nature of manual usage compared to the journals. In fact, all readers specifically remarked that manuals were only ever used in work or 'task' domains.

What type of information is in a text?

THE TEXT/GRAPHICS DICHOTOMY

A major distinction seems to be drawn by readers between text and graphics in describing information types. The general consensus was that academic journals are a predominantly textual rather than graphical form of documentation (mentioned by 60 per cent) while all readers invoked this dichotomy in reporting that manuals relied too heavily on textual contents. Graphics in articles (for example, tables, figures, etc.) were generally viewed positively as seven readers explicitly stated a dislike for academic articles that consisted of pages of straight text. A high proportion of mathematical content was also viewed negatively by the readers in this sample. More graphics would often aid the manual user's location and comprehension of information it was observed. However, it was repeatedly pointed out that graphics should be relevant and one manual was much criticised for using superfluous pictures of desktops, scissors and documents.

LINGUISTIC STYLE

The language style was also mentioned in this context. The language of academic journals was seen to be relatively technical such that only readers versed in the subject matter could profitably read it. Furthermore, presentation style is seen as highly formalised, that is, material written in a manner unique to journals that differs from conventional prose.

In the case of software manuals, material was invariably described as 'technical', 'specific' and 'detailed'. While it might be argued that the very nature of manuals is that they contain such information, most readers seemed to find this off-putting. A third of the readers remarked that manuals were heavily loaded with jargon, with the result that information on simple actions was often difficult to locate or extract.

INFORMATION STRUCTURE

The concept of structure in articles was discussed with all readers and it was apparent from their responses that most readers are aware of this as an organising principle in documents. Organisation of academic journal articles around a relatively standard framework was articulated, which the majority of the present respondents viewed as being of the form:

- Introduction
- Method
- Results
- Discussion/Conclusion

or of the form:

- Introduction
- Elaboration and criticism of issues
- Alternative views proposed by author
- Discussion/Conclusion

depending on whether it is an experimental or review type paper.

This order was seen as useful for reading purposes as it facilitated identification of relevant sections and allowed rapid decision making on the suitability of the article to a reader's needs (a hypothesis that we went on to test formally, see next chapter). For example, poor sectioning, large method and results sections, small discussions and large size in terms of number of pages were all cited as factors that would influence a reader's decision on whether or not to reject an article.

The structure of manuals was discussed with all readers and responses varied between those who are aware of text structure as it pertains to this text type and those who felt it existed but had difficulty articulating their perceptions of it. Primarily, a sense of order seems to be lacking in manuals though the majority (60 per cent) of readers felt that there might be a progression from easy to hard as a reader moves from the beginning to the end of the text; that is, the more complex operations are dealt with towards the back of the manual. One reader remarked that a structure based around command frequency was probably common; that is, frequently used commands or actions were more likely to be located at the front of the manual and less common ones at the back. Another suggested order, general-to-specific details, was made by two readers. Two readers argued that if any order such as easy-to-hard could be observed it probably existed at the task rather than the global level, that is, within sections rather than across the manual.

The perceived modal structure for manuals that emerged in this study was:

- contents
- getting started
- simple tasks
- more complex tasks
- index.

As this structure indicates, heavy emphasis was placed on the task as a structural unit for organising manuals. There were variations on this modal structure. For example, two readers placed training exercises at various points in the structure, the gradation between basic and more complex tasks was extended in two cases to include an intermediate level, while others mentioned a glossary of commands, technical specifications and lists of error messages as further typical units of a manual.

Many of the problems users of manuals seem to experience are related to the question of structure. Invariably this was criticised as 'poor' or 'disorganised'. The present sample seemed divided between those who felt that overall order was less important than the procedural order at the task level and those who were content with procedural ordering but felt that high-level ordering was unsatisfactory in many manuals. The need for different versions of manuals which are structured according to the users' needs was suggested by four readers. Typically it was suggested that these should consist of a manual for a 'total novice' which explains how to perform very basic procedures and a more detailed version for users who have acquired a greater degree of competence.

DOCUMENT SIZE

The issue of document size is also interesting. Reading lengthy academic articles obviously require a significant time-investment, which is often seen as a disincentive. Perceptions of what constituted a large or small article varied. Large articles were described as being anything from six to more than 30 pages long, medium-length articles as being between five and 20 pages long and small articles as being between three and 20 pages long. In other words, what one individual rates as large, another may rate as small. Median responses suggest that articles more than 20 pages long are large and those articles that are about five pages long are small. Approximately 10 pages are considered to be medium length.

How are texts read?

Figure 6.1 represents a generic description of academic journal usage patterns. First, all readers skim read the table of contents of the journal issue. A preference was expressed for contents lists printed on the front or back page, which made location of relevant articles possible without opening the journal. If the reader fails to identify anything of interest at this point the journal is put aside and, depending on the circumstances, further journals may be accessed and their contents viewed as above. When an article of interest is identified then the reader opens the journal at the start of the relevant paper. The abstract is usually attended to and a decision made about the suitability of the article for the reader's purposes.

At this point most readers reported also browsing the start of the introduction before flicking through the article to get a better impression of the contents. Here readers reported attending to the section headings, the diagrams and tables, noting both the level of mathematical content and the length of the article. Browsing the conclusions also seems to be a common method of extracting central ideas from the article and deciding on its relevance.

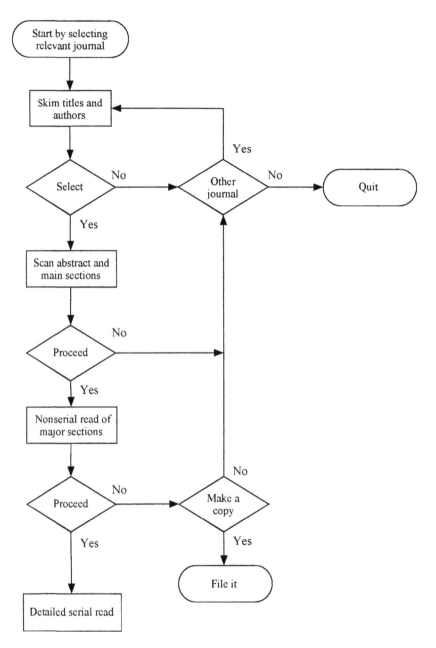

Figure 6.1 Generic model of journal usage.

When readers have completed one initial cycle of interaction with the article they make a decision whether or not to proceed with it. A number of factors may lead to the reader rejecting the article. The main reason is obviously content and the accuracy of this first impression is an interesting empirical question.

If the article is accepted (or, more likely, photocopied) for reading it is likely to be subjected to two types of reading strategy. The rapid scan reading of the article, usually in a nonserial fashion to extract relevant information, involves reading some sections fully and only skimming or even skipping other sections. The second reading strategy seemed to be a serial detailed read from start to finish. This was seen as 'studying' the article's contents and though not carried out for each article that is selected, most readers reported that they usually read selected articles at this level of detail eventually. Three readers expressed a preference for this reading strategy from the outset over scanning though acknowledging it to be less than optimum. While individual preferences for either strategy were reported, most readers seem to use both strategies depending on the task or purpose for reading the article, time available and the content of the article.

Figure 6.2 represents usage styles for software manuals in a similar flowchart form. The first thing readers reported is 'getting a feel' for the document's contents. Thus readers initially open the text at the contents or index sections. The majority (60 per cent) stated that the contents page is usually checked first and if that did not suggest where to go, the index was examined. However, it seems that much depends on the nature of the problem. If decoding an error message or seeking information on a particular command then the index is likely to provide this and will therefore be accessed first.

If more general information is sought then the contents offer a better chance of location. This highlights the extent to which book conventions have become internalised in the minds of contemporary readers. Furthermore, the index or contents list is not read in a simple pattern-matching fashion for a word template; rather the reader tries to get an impression of context from the contents, that is, what items precede and proceed it, or may have to think of other terms that might yield satisfactory information if the term being sought is not present (a common problem with technical jargon).

If the reader fails to locate anything in the contents or index that appears relevant he may either dip into the text and move about looking for relevant information or give up. The latter option appears common according to the data and represents a failure in document design that must be overcome. If, on the other hand, a relevant item is located then the reader turns to the relevant page and if it is not immediately obvious, will scan about looking for a relevant diagram, word, phrase, etc., to indicate that the answer is there.

At this stage the reading strategy adopted by the reader depends on the particulars of the task. The tendency to 'get-on-with-it' seems firmly established in users of manuals and the present sample report moving freely

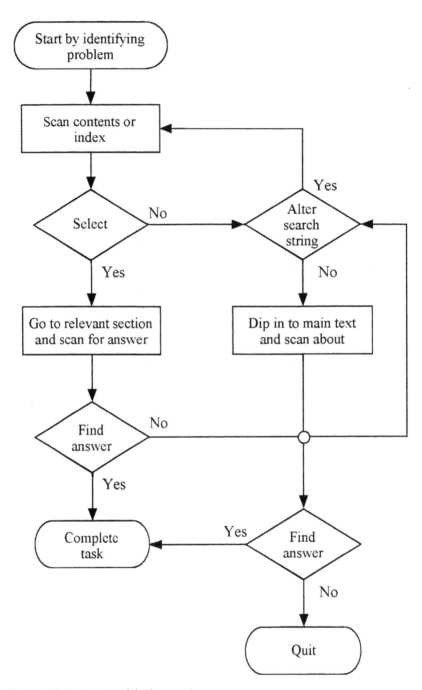

Figure 6.2 Generic model of manual usage.

from manual to system in order to achieve their goal. Only three readers manifested any tendency to read around an area or fully read a section before moving on and even these admitted that they would be tempted to skim, tend to get bored if they felt that they were not resolving their problems and only read complete sections if all else had failed.

Summarising the usage data

In terms of the situated simulation approach followed here, the texts can be summarised as follows:

Academic journals

Why: For work reasons such as keeping up with the literature, as a source of reference and as a source of learning. They are also read for personal reasons when accessed out of interest with no immediate work requirements.

What: Technical information about a specific domain; may have graphical components but are predominantly textual and tend to conform to a relatively standard structure.

How: Three levels of reading:

(i)   quick scan of abstract and major headings;
(ii)   nonserial scan of major sections;
(iii)   full serial read of the text.

Software manuals

Why: For task-specific reasons such as troubleshooting, getting started, and for reference. Can occasionally be used for exploring software and identifying facilities or obtaining detailed technical information.

What: Technical information, of a specific and detailed nature, often laden with jargon. Can be a mixture of text and graphics. Structure is based around task units.

How: Problem driven. Broadly it involves checking the index or contents sections to find something relevant, then dipping into and scanning sections of text. Lengthy serial reading is rare.

Thus, it is possible to gain information on the reading process for various texts using this approach and this information is more reliable than would be gained merely by self-analysis of one's own experiences with a document.

One should obviously place these data in context, by which I mean that one should make explicit, along with the Why/What/How themes, *where* the information source will be used, *when* it will be used and by *whom*. I do not formally mention these issues in this method since such contextual variables are part of the standard usability approach outlined in Chapter 2, which requires the context of use to be articulated explicitly in the operationalisation of usability. The What/Why/How issues are presented as further issues to explore only after the context of use has been established (see Chapter 2 for a more detailed discussion of usability and the measurement of context).

Conclusions and the way forward

Though probably a nonoptimum means of gathering reading process data, this simulation-based procedure demonstrates that it is at least possible to gain design-relevant information on the reading process without recourse to sophisticated and/or unnecessarily intrusive tools. It has been my experience that readers have few difficulties articulating how they use texts and responding to questions on the What and Why aspects of reading. There is normally a high degree of consistency between readers, which enables general conclusions to be drawn, yet the data clearly distinguish between text types in the characterisation of usage they suggest. Furthermore, it can be applied to both paper and digital texts.

In an extension of this method, Vaughan (1999) used it as part of her requirements gathering phase for the design of digital newspapers. In this case she employed it in a group setting to solicit information from experienced users and used other data gatherers to help her capture all the data. Once again, the method is flexible enough to be modified for use as needed.

Data of this type are useful for gaining insight into the relevant design issues for an electronic document, particularly early in the product life cycle (a time when human factors inputs are typically weak) but there are limitations. It is applicable, in the main, only for existing texts. For innovative information types such a classification of task-relevant issues is not so simple (although subsets of the hypermedia information would be amenable to such analysis, and users could always speculate on probable or likely Whys and Whats). Furthermore, it is a relatively informal procedure, reliant for its value ultimately on the abilities of the practitioner more than one might like, particularly for mapping responses to interface recommendations.

However, this form of data capture and analysis is not intended to be anything more than a means of initially conceptualising the issues, that is, identifying the text type and its associated usage characteristics. By providing designers with some reliable estimates of the three aspects of usage – the why, what and how of reading – it supports reasoned constraining of the design options under consideration. More formal analyses can then be appropriately targeted at specific design issues for the artifact under consideration.

As an investigative procedure, it is quick to perform and seems unlikely to tax readers unduly. For sure it does not reveal where the reader's eyes go or accurately map the sequence of pages followed but, in defense, one must ask if that is what is needed at this stage? Approximate data of these kinds offer a means of constraining the space of design options and highlighting aspects of reading that the technology could improve. In so doing it involves intended users of the technology and offers a means of negotiating potential solutions with them – a requirement of any truly user-centred approach.

So we can gain much from users early on with such methods, but one issue that appears difficult to tackle directly through observation or inter-viewing is the structure of information space, which underlies the navigation issue. It is to this topic that attention is turned in the next chapter.

7 Shape: information as a structured space

Order is the shape upon which beauty depends.

Pearl Buck, 1966

Introduction

The emergence of new electronic document forms makes it possible to embody alternative and multiple structures for electronic texts that could not be supported feasibly in the comparatively standard format of paper. Typically, advocates of the 'new structures' approach dismiss paper as a limiting medium, claiming it imposes a 'linear straitjacket of ink on paper' (Collier 1987). This is contrasted with the supposedly liberating characteristics of hypertext, which, with its basic form of nodes and links, is seen to be somehow freer or more natural. Nielsen (1990), for example, would appear to reflect the consensus view with the following summary:

> All traditional text, whether in printed form or computer files, is sequential, meaning that there is a single linear sequence defining the order in which it is to be read. First you read page one. Then you read page two. Then you read page three. And you don't have to be much of a mathematician to generalize the formula which determines which page to read next . . . Hypertext is nonsequential; there is no single order in which the text is to be read.
>
> (p. 1)

Similar and even more extreme positions can be found in many of the (largely paper-based) writings describing the wonders of hypertext (e.g., Landow 1992; Dryden 1994).

One may not need to be a sophisticated mathematician to derive a formula to predict such a reading style but one would be a poor student of human psychology if one really believed any resulting formula provided an accurate representation of the reading process. Certainly, reading a sentence is largely a linear activity (although eye-movement data suggest that, even here, nonlinearity can occur) so even on screen, reading of sentences is still as likely to be predominantly linear. The issue then is the extent to which linearity is imposed on the reader at the suprasentence level in using the document.

Although paper texts may be said to have a physical linear format there is little evidence to suggest that readers are constrained by this or only read such texts in a straightforward start-to-finish manner (Pugh 1975; Charney 1987; Horney 1993). The journal usage study in the previous chapter identified three reading strategies in readers of academic journals, only one of which could be described as linear, and one only has to think of one's own typical interactions with a newspaper to demolish arguments of constrained linear access and talk of 'straitjackets.' With digital documents, movement might involve less physical manipulation (a keypress, for example, compared to turning and folding pages) or less cognitive effort (a direct link to information compared to looking up an index and page number) but these are matters of degree, not of category. One could make a case for paper being the liberator as at least the reader always has access to the full text (even if searching it might prove awkward). With digital documents, the absence of links could deny some readers access to information and always force them to follow someone else's ideas of where the information trail should lead.

Although the notion of linearity is (to my mind) much abused in this domain, the notions of access paths and structure have given rise to the idea of information as having shape and occupying space through which a reader must navigate. In seeking to provide innovative and/or natural structures for information, the question of navigation has become a central one to the field. In the present chapter, these ideas are examined and the extent to which text designers must address such issues is considered.

Tangibility and the concept of structure in documents

Regardless of the putative constraints of paper texts or browser-friendly attributes of hypermedia, it seems certain that on approaching a document, readers possess some knowledge of it that provides information on the probable structure and organisation of key elements within it. For example, the moment we pick up a book we are afforded numerous cues that as experienced readers we are likely to respond to, such as size (which can be an indicator of likely effort involved in reading it), age or condition (which can act as an indicator of relevance and also of previous usage rates). Such affordances give at least partial lie to the cliché that you cannot judge a book by its cover; you can, and you do.

Once the book is opened, still more cues are available to readers, informing them of where material is located, how it is organised and what information is included. This information is available through contents lists, chapter headings, abstracts and summaries; for some this is termed the 'architecture' of the space.[1] The regular reader comes to expect this information and probably picks it up in a relatively automatic fashion when browsing a book. Almost certainly, we would think it odd if the contents list was missing or the index was at the front. We expect certain consistency in layout and structure and notice them most when this expectation is violated.

The same point might be argued for a newspaper. Typically we might expect a section on the previous day's political news at home, foreign coverage, market developments and so forth. News of sport will be grouped together in a distinct section and there will also be a section covering that evening's television and radio schedules. With many paper documents there tend to be at least some standards in terms of organisation so that concepts of relative position in the text such as 'before' and 'after' have tangible physical correlates. If this can be said to hold true for all established text forms, then developers of electronic systems need to consider carefully their designs in terms of whether they support or violate such assumptions. Such regularities are emerging also in their own way for digital documents too, albeit superficially; just consider the placement of navigation bars on most web pages, for example.

Unfortunately, the term 'structure' is used in at least three distinct ways by different researchers and writers in this field. Conklin (1987) talks of structure being imposed on what is browsed by the reader, that is, the reader builds a structure to gain knowledge from the document. Trigg and Suchman (1989) refer to structure as a representation of convention, that is, it occurs in a text form according to the expected rules a writer follows during document production. Hammond and Allinson (1989) offer a third perspective, that of the structure as a conveyer of context. For them, there is a naturally occurring structure to any subject matter that holds together the 'raw data' of that domain and supports reading.

In reality, these are manifestations of the same concept, for which I prefer to use the word 'shape'. The main role of shape seems to differ according to the perspective from which it is being discussed, the writer's or the reader's, and the particular part of the reading/writing task being considered. Thus the shape of a document can be a convention to both the writer, so that she conforms to expectations of format, and the reader, so she knows what to expect. It can be a conveyer of context mainly to the reader so she can infer from, and elaborate on, the information provided, but it might be employed by a skilled writer with the intention of provoking a particular response in the reader. Finally, it can be a means of mentally representing the contents to both the reader, so she grasps the organisation of the text, and the author, so that she can appropriately order this delivery.

---

1   This is not really the place to go into the 'little IA' versus 'big IA' debate, you can find more of this in Dillon (2002). However, if pushed, such a description would be a 'little' IA usage of the term 'architecture'.

van Dijk and Kintsch's (1983) theory of discourse comprehension places great emphasis on text structure in skilled reading. According to their theory, reading involves the analysis of propositions in a text and the subsequent formation of a macropropositional hierarchy (that is, an organised set of global or thematic units about the events, acts, and actors in the text). From this perspective, increased experience with texts leads to the acquisition of knowledge about regularities which van Dijk and Kintsch term 'superstructures' that facilitate comprehension of material by allowing readers to predict the likely ordering and grouping of constituent elements of a body of text in advance of reading it.

They have applied this theory to several text types. For example, with respect to newspaper articles they describe a schema consisting of headlines and leads (which together provide a summary), major event categories, each of which is placed within a context (actual or historical), and consequences. Depending on the type of newspaper (for example, weekly as opposed to daily, tabloid as opposed to quality, etc.), one might expect elaborated commentaries and evaluations. Experiments by Kintsch and Yarborough (1982) showed that articles written in a way that adhered to this schema resulted in better grasp of the main ideas and subject matter (as assessed by written question answering) than ones which were reorganised to make them less schema conforming.

Interestingly, when given a cloze test (a traditional comprehension test for readers that requires them to fill in the blanks within sentences taken from the text they have just read) of the articles, no significant difference was observed. The authors explain this finding by suggesting that schematic structures are not particularly relevant as far as the ability to remember specific details such as words is concerned (that is, the ability which is measured by a cloze test) but have major importance at the level of comprehension. In their terms, word processing and recall is handled at the microstructural level, text-specific organisation at the macrostructural level and general organisation of the text type at the superstructural level. We have seen similar results in our tests of genre-conforming and genre-violating digital newspapers also, where the advantages of form can be seen at the general comprehension level, not at the specific detail level (see Vaughan 1999 for details).

The van Dijk and Kintsch theory has been the subject of criticism from some cognitive scientists. Johnson-Laird (1983), for example, takes exception to the idea of any propositional analysis providing the reader with both the basic meaning of the words in the text and the significance of its full contents. For him, at least two types of representational format are required to do this and he provides evidence from studies of people's recall of text passages that it is not enough to read a text correctly (that is, perform an accurate propositional analysis) to appreciate the significance of that material. He proposes what he terms mental models as a further level of representation that facilitates such understanding. Subsequent work by Garnham (2001) lends support to the insufficiency-of-propositions argument in comprehension of text.

The differences between Johnson-Laird and van Dijk are mainly a reflection of the theoretical differences between the psychologist's and the linguist's views of how people comprehend discourse. From the perspective of the human factors practitioner it is not clear that either theory of representation format is likely to lead to distinct (that is, unique) predictions about electronic text design. Both propose that some form of structural representation occurs – it is just the underlying cognitive form of this representation that is debated.

The debate on representational format largely sidesteps the structural issue of most relevance to electronic text design, the high-level organisation of presented material. This is van Dijk and Kintsch's third level, the super-structural, which is akin to the text conventions and organisation principles. As yet, we have had too little theoretical debate at this level. For the user experience practitioner, theoretical distinctions of supposed representational form are unlikely to be as important to the design of electronic documents as consideration of the physical text structure that gives rise to the cognitive experience and the concomitant so-called 'representations'. The major issue in this context, therefore, is the extent to which structure is perceived in certain texts and how designers must accommodate this in their products.

Schema theory as an explanatory framework for information structure

From the comments of readers in our studies it is clear that shape or structure is a concept for which the meanings described seem to apply with varying degrees of relevance, though many often find it difficult to articulate. So how might we explain the perception of shape?

Viewing structure as a component of texts leads directly to the view of information as space and the reader as a navigator. This in turn invites a direct mapping between psychological theory and information design that has been unquestioningly accepted by researchers in this domain.

It seems obvious, for example, that we must possess schemata of the physical environment we find ourselves in if we are not to be overwhelmed by every new place we encounter. Presumably acquired from exposure to the world around us, schemata afford a basic orienting frame of reference to the individual. Thus, we soon acquire schemata of towns and cities so that we know what to expect when we find ourselves in one: busy roads, numerous buildings, shopping areas, people, etc.

In employing schema theory as our explanatory framework it is worth making a distinction between what Brewer (1987) terms 'global' and 'instantiated' schemata. The global schema is the basic or raw knowledge structure. Highly general, it does not reflect the specific details of any object or event (or whatever knowledge type is involved). The instantiated schema, however, is the product of adding specific details to a global schema and thereby reducing its generality. An example will make this clearer. In orienting ourselves in a new environment we call on one or more global schemata (for example, the schema for city or office building). As we proceed to relate specific details of our new environment to this schema we can be said to develop an instantiated schema which is no longer general but is not sufficiently complete to be a model or map of the particular environment in which we find ourselves.

Global schemata remain to be used again as necessary. Instantiated schemata presumably develop in detail until they cease to be accurately described as schematic or are discarded when they serve no further purpose. In the above example, if we leave this environment after a short visit we are likely to discard the instantiated schema we formed, but if we stay or regularly return we are likely to build on this until we have detailed knowledge of the environment.

While schemata are effective orienting guides, in themselves they are limited. In particular, they fail to reflect specific instances of any one environment and provide no knowledge of what exists outside of our field of vision. As such, they provide the basic knowledge needed to interact with an environment but must be supplemented or instantiated by other representations if we are to plan routes, avoid becoming lost or identify short cuts – activities in which humans seem frequently to engage.

Levels of schema instantiation: landmarks, routes and surveys

The first postulated stage of schema instantiation is knowledge of landmarks, a term used to describe any features of the environment which are relatively stable and conspicuous. Thus we recognise our position in terms relative to these landmarks; for example, our destination is near a large, new building or if we see a statue of a soldier on horseback then we must be near the railway station and so forth. This knowledge provides us with the skeletal framework on which we build our cognitive map.

The second stage of instantiation is route knowledge which is characterised by the ability to navigate from point A to point B, using whatever landmark knowledge we have acquired to make decisions about when to turn left or right. With such knowledge we can provide others with effective route guidance; for example, 'Turn left at the traffic lights and continue on that road until you see the large church on your left and take the next right there . . . ' and so forth. Though possessing route knowledge a person may still not really know much about her environment since any given route might be effective but nonoptimal or even totally wasteful.

The third stage of instantiation is survey (or map) knowledge. This allows us to give directions or plan journeys along routes we have not directly travelled as well as describe relative locations of landmarks within an environment. It allows us to know the general direction of places; for example, 'westward' or 'over there' rather than 'left of the main road' or 'to the right of the church'. In other words it is based on a world frame of reference rather than an egocentric one.

Current thinking is dominated by the view that landmark, route and survey knowledge are points on a continuum rather than discrete forms. The assumption is that each successive stage represents a developmental advance towards an increasingly accurate or sophisticated world view. Certainly this is an intuitively appealing account of our own experiences when coming to terms with a new environment or comparing our knowledge of one place with another and has obvious parallels with the psychological literature which often assumes invariant stages in cognitive development, but it might not be so straightforward.

Obviously, landmark knowledge on its own is of little use for complex navigation and both route and survey knowledge emerge from it as a means of coping with the complexity of the environment. However, it does not necessarily follow that given two landmarks the next stage of knowledge development is acquiring the route between them or that once enough route knowledge is acquired it is replaced by or can be formed into survey knowledge. Experimental investigations have demonstrated that each form of representation is optimally suited for different kinds of tasks. Route knowledge is cognitively simpler than survey knowledge but suffers the drawback of being more or less useless once a wrong step is taken (Wickens 1991). Thus while the knowledge forms outlined here are best seen as points on a continuum and a general trend to move from landmark to survey knowledge via route knowledge may exist, task dependencies and cognitive ability factors mediate such developments and suggest that an invariant stage model may not be the best conceptualisation of the findings.

Can we consider documents as navigable structures?

If picking up a new book can be compared to a stranger entering a new town (that is, we know what each is like on the basis of previous experience and have expectancies of what we will find even if we do not yet know the precise location of any element), how do we proceed to develop our map of the (other than physical) information space?

To use the analogy of the navigation in physical space we would expect that generic structures such as indices, contents, chapter headings and summaries can be seen as landmarks that provide readers with information on where they are in a text, just as signposts, buildings and street names aid navigation in physical environments. Thus when initially reading a text we might notice that there are numerous figures and diagrams in certain sections, none in others, or that a very important point or detail is raised in a section containing a table of numerical values. In fact, readers often claim to experience such a sense of knowing where an item of information occurred in the body of the text even if they cannot recall that item precisely and there is some empirical evidence to suggest that this is in fact the case.

Rothkopf (1971) tested whether such occurrences had a basis in reality rather than resulting from popular myth supported by chance success. He asked people to read a 12-page extract from a book with the intention of answering questions on content afterwards. What subjects did not realise was that they would be asked to recall the location of information in the text in terms of its occurrence within both the page and the complete text. The results showed that incidental memory for locations within any page and within the text as a whole were more accurate than chance, that is people could remember location information even though they were not asked to. There was also a positive correlation between location of information at the within-page level and accuracy of question answering. Work by Zechmeister *et al.* (1975) and Lovelace and Southall (1983) confirm the view that memory for spatial location within a body of text is reliable even if it is generally limited. Simpson (1990) has replicated this finding for electronic documents of a nonlinked variety.

In the paper domain at least, the analogy with navigation in a physical environment is of limited applicability beyond the level of landmark knowledge. Given the fact that the information space is instantly accessible to the reader (that is, she can open a text at any point), the necessity for route knowledge, for example, is lessened (if not eliminated). To get from point A to point B in a text is not dependent on taking the correct course in the same way that it is in a physical three-dimensional environment. The reader can jump ahead (or back), guess, use the index or contents or just page serially through. Readers rarely rely on just one route or get confused if they have to start from a different point in the text to go to the desired location, as would be the case if route knowledge was a formal stage in their development of navigational knowledge for texts. Once you know the page number of an item you can get there as you like. Making an error is not as costly as it is in the physical world either in terms of time or effort. Furthermore, few texts are used in such a way as to require that level or type of knowledge.

One notable exception to this might be the knowledge involved in navigating texts such as software manuals or encyclopaedias, which can consist of highly structured information chunks that are interreferenced. If, for example, a procedure for performing a task references another part of the text, it is conceivable that a reader may only be able to locate the referenced material by finding the section that references it first (perhaps because the index is poor or she cannot remember what it is called). In this instance one could interpret the navigation knowledge as being a form of route knowledge. However, such knowledge is presumably rare except where it is specifically designed into a document as a means of aiding navigation along a troubleshooting path.

A similar case can be made with respect to survey knowledge. While it seems likely that readers experienced with a certain text can mentally envisage where information is in the body of the text, what cross-references are relevant to their purpose and so forth, we must be careful that we are still talking of navigation and not changing the level of discourse to how the argument is developed in the text or the ordering in which points are made. Without doubt, such knowledge exists, but often it is not purely navigational knowledge but an instantiation of several schemata such as domain knowledge of the subject matter, interpretation of the author's argument, and a sense of how this knowledge is organised that come into play now. This is not to say that readers cannot possess survey-type knowledge of a text's contents, rather it is to highlight the limitations of directly mapping concepts from one domain to another on the basis of terminology alone. Just because we use the term navigation in both situations does not mean that they are identical activities with similar patterns of development. The simple differences in applying findings from a three-dimensional world (with visual, olfactory, auditory and powerful tactile stimuli) to a two-dimensional text (with visual and limited tactile stimuli only) and the varying purposes to which such knowledge is put in either domain are bound to have a limiting effect.

It might be that rather than route and survey knowledge, a reader develops a more elaborated analogue model of the text based on the skeletal framework of landmark knowledge outlined earlier. Thus, as familiarity with the text grows, the reader becomes more familiar with the various landmarks in the text and their interrelationships. In effect the reader builds a representation of the text similar to the survey knowledge of physical environments without any intermediary route knowledge but in a form that is directly representative of the text rather than physical domain. This is an interesting empirical question and one that is far from being answered by current knowledge of the process of reading.

Navigable structures and electronic space

The concept of a schema for an electronic information space is less clear-cut than for physical environments or paper documents. Computing technology's relatively short history is one of the reasons for this but it is also the case that the medium's underlying structures do not have equivalent transparency. With paper, once the basic *modus operandi* of reading is acquired (for example, page-turning, footnote identification, index usage and so forth), it retains utility for other texts produced by other publishers, other authors and for other domains. With computers, manipulation of information can differ from application to application within the same computer, from computer to computer and from this year's to last year's model. Thus using electronic information is often likely to involve the employment of schemata for systems in general (that is, how to operate them) in a way that is not essential for paper-based information.

The qualitative differences between the schemata for paper and electronic documents can easily be appreciated by considering what you can tell about either at first glance. When we open a digital document, we do not have the same amount of information available to us. We are likely to be faced with a welcoming screen which might give us a rough idea of the contents (that is, subject matter) and information about the authors/developers of the document but little else. It is two-dimensional, gives no indication of size, quality of contents, age (unless explicitly stated) or how frequently it has been used (that is, there is no dust or signs of wear and tear on it such as grubby finger-marks or underlines and scribbled comments). At the electronic document level, there is usually no way of telling even the relative size without performing a query, which will usually return only a size in kilobytes which conveys little meaning to the average reader.

The fact that digital information offers authors the chance to create numerous structures out of the same material is a further source of difficulty for users or readers. Since schemata are generic abstractions representing typicality in entities or events, the increased variance of digital documents implies that any similarities that are perceived must be at a higher level or must be more numerous than the schemata that exist for paper texts. It seems, therefore, that users' schemata of such environments are likely to be 'informationally leaner' than those for paper documents. This is attributable to the recent emergence of electronic documents and comparative lack of experience interacting with them as opposed to paper texts for even the most dedicated users. The current lack of standards in the electronic domain compared to the rather traditional structures of many paper documents is a further problem for schema development.

Beyond navigation: the spatial-semantics of information shape

Navigation in information spaces is certainly a major source of cognitive overhead for users of digital information systems. Frequently, the overhead is such that disorientation is experienced (the 'lost in hyperspace' phenomenon) and users have difficulty locating required information. At best, this problem leads to increased time taken to cover material or attain acceptable comprehension rates with hypermedia (see for example, McKnight *et al.* 1990); at worst, it can lead to the rejection of this technology on the part of users.

Plausibly, disorientation can occur through the overloading of short-term memory (STM) as users are required to remember their paths or attend to spatial markers. Overcoming STM limitations through signs and back-tracking facilities is a practical strategy directly under the control of the design team. Thus we have an extensive literature on the value and applicability of such spatial components of interfaces as menus (Norman 1991), maps and browsers (Simpson and McKnight 1990), pop-up windows (Stark 1990), direct jumping (Wright and Lickorish 1990), etc. Such findings can work well where our goal is to support movement through an information space to a specific location, although it is cautionary that emerging datasets of user behaviour in web environments reveal a greater reliance on simple 'back' button navigation rather than the designed history feature of a browser (Byrne *et al.* 1999). This feature-oriented approach to research and design derives from the classic journey metaphor of hypermedia use (see, for example, an early manifestation of this journey metaphor in Hammond and Allinson 1987 and a current manifestation of it in Calvi 1997).

Over the past few years I have argued that this navigation model and its attendant focus on screen features bypasses discussions of semantic space. Yet typical users are often directly interacting with the meaning of information, not consciously seeking to navigate through a space. Thus, to ensure better design for learning and communication, greater emphasis should be placed on the semantic issues that impact use. Such arguments have been extended in recent years to the notion of information possessing shape (those spatial-semantic properties that convey coherence) that users can exploit both semantically and physically to gather meaning (Dillon and Schaap 1996).

The concept of shape assumes that an information space of any size has both spatial and semantic characteristics. That is, as well as identifying placement and layout, users directly recognise and respond to content and meaning. Routinely in our laboratory, users describe what they remember from an interaction in digital space or draw their interpretation of the information space's form and layout. These data clearly point to the inter-coupling of spatial and semantic components of memory. For example, when asked to describe an information space after interaction, users employ terms that convey relationships and elaborations as well as purely spatial linkages such as position and sequence (see for example, Dillon and Vaughan 1997). Completely separating both forms of representation is rare and somewhat artificial to users of an information space. Users easily move from one to the other since both serve to advance their desire for task completion. Indeed, it makes best sense to think of the user's model of the information space as being constructed out of both.

Such studies suggest that while location of information is necessary, it is unlikely that location itself will be sufficient in many interactive tasks. Once found, information is processed for relevance and meaning to the user. Reactions to this information cause further interactive behaviours. As future technologies seek to immerse users in a three-dimensional information space of multiple documents and clusters of information to enhance learning, knowledge creation and information transfer, a design target of 'supporting location through navigation' will likely prove overly limited.

Futurism notwithstanding, there is a compelling reason to examine the semantic issues of use. Humans manifest a native cognitive tendency to impose structure on information through use which is crucial to identifying appropriate information visualisations. 'Shaping displays', those forms of presentation that represent spatial and semantic properties in meaningful forms, are essential if we are to move the technology from its current existence as an access mechanism to its future as a knowledge tool or an augmenter as originally envisaged by Bush, Engelbart, and Nelson.

Examining the spatial-semantics of information shapes

In a series of studies exploring user's abilities to locate themselves in electronic text environments, my students and I have shown repeatedly that frequent users become attuned to regularities in form and expression of ideas that distinguish certain elements of an information space in semantic terms, and these may override spatial cues in providing users with a sense of location. For example, in a series of studies of users' ability to categorise isolated paragraphs of a scientific article presented on screen, Dillon (1991) originally showed that experts could perform this task with 80 per cent accuracy without reading for comprehension. These text items were presented with no obvious spatial information available to the reader, thus experts must have been able to base their location on implicit signs in the discourse. It was assumed that there must exist clear perceptual cues in the discourse (keywords, formulae, etc.) that experienced readers can exploit easily to determine location since such cues might represent specific examples of categorical form.

Dillon and Schaap (1996) then manipulated the presence or absence of textual cues and examined the role of expertise with this discourse form in performing such location tasks. Results showed a significant effect for expertise, with novices manifesting greater numbers of errors than experts in terms of information location. The authors concluded that the ability to sense where one is in information space seems to be reliant only in part on the rapid identification of visual details present in the information space. Experts exploit more than visual information to gain a clear sense of location, they also apply semantic processing and thus can categorise the information display as belonging to a certain part of a larger information space or structure. Novices or inexperienced users of an information type do not possess the necessary knowledge to interpret these cues and therefore must rely on explicit spatial indicators (such as headings or titles) in the visual display or on their own limited knowledge structure of the general class of information types in the world.

However, the precise form of cue that experts latched onto was not simply isolated. Dillon and Shaap (1996) made a plausible case for experimentally manipulating obvious (to experts) visual cues or signs such as statistical formulae, author references, test results, descriptions of methods, etc. As such, these could be seen as representing basic semantic cues inherent in the language of this information space which are invariant across the form. Interestingly, in this study the cue manipulation was not a significant major effect, rather it interacted significantly with category (that is, superstructural form of the information space). The implication from this is that the cues manipulated were insufficient to explain the basis for the experts' superior performance.

If experts were developing a sense of location out of more than the primitive visual cues varied in these presentations then it is clear that some nonsurface-level representation in the display was affording them information on the organisation of the document. To identify what these might be we examined the verbal protocols of experts as they attempted this task. In a further investigation the presence or absence of the cues originally manipulated by Dillon and Schaap (1996) were counterbalanced across categories. The subject's task at this point was to categorize each paragraph and state the reasons verbally for their choice of location.

The verbal protocols from this test are interesting. Consider for example the following text which was used as a stimulus:

> In addition to assessing more completely the entire problem-solving process, the modified problem-solving measure has another advantage over the problem-solving measure; rather than using contrived hypothetical problems and solutions to assess problem-solving skills, it is adaptable to any idiosyncratic problem. The ability of an individual to deal with specific problems can therefore be assessed. In the present study, we asked subjects to problem-solve their response to obtaining a poor grade on an introductory psychology exam prior to obtaining such a grade. Problem-solving deficits regarding this specific stressor were expected to predict response to this stressor.

Asked to categorize this paragraph (which is an introduction paragraph from a scientific article, stripped of obvious cues), experts presented protocols such as these:

> 'Well this is interesting. My initial impression was that this is going to be discussion. And then I shifted back to methods. It's talking mostly about . . . measures. Um . . . it could be . . . No it's introduction actually. It's because it's an overview of the method. Some studies put this in an overview of a methods section . . . Most likely here it's the introduction, and the other thing that indicates that we're not further on, say in discussion is that it's talking about what's expected to happen . . . '

From another expert:

> 'Well, I am stuck . . . Don't know where to put this one. It's either in an intro where you are setting up what you are about to tell me in more

depth or you've already told me in depth and now you are presenting it to me in a discussion – so that's where I'll put it, I can see it in either. But probably . . . probably more in the discussion because it seems as if I'd know more about what this 'modified problem solving measure' was . . . You would have talked about it more at length, which was earlier, So this is discussion.'

So even experts disagree, particularly for introduction and discussion sections which were previously identified as the most common mistake in such location tasks. However, without explicit cues, experts seek to base their decisions on content ('it's talking about measures') and on logically anticipated form of the argument flow – 'it's talking about what's expected to happen' (example 1) and 'you would have talked about it more at length' (example 2). In both cases we see an appeal to the inherent structure of the form or genre made by users who are expert in this discourse type without any explicit cueing in the presentation. Such users are navigating through an information space but not in a way that is equivalent to a physical journey through space. For these users it is more a case of abstracting structure from the match between presented information and generic form which they have internalised over a lengthy socialisation and training period.

It would appear that experts are capable of exploiting knowledge shared across the community of practice. Novices are left to logically infer everything from the language without any reference to expected form or structure. While it seems useful to design explicit spatial representations to advance the novice user's performance on such tasks, a more urgent design goal is to determine how a novice might gain greater semantic knowledge of the content through the representations we design. If, as van Dijk and Kintsch (1983) have long argued, information has a form that reflects its community's practices, we may find that designing the information space to take account of the shaping process has commensurate benefits in training new practitioners in a discipline to construct meaning.

The spatial-semantic model of information shape perception

My work since the first edition of this book suggests that the interplay of implicit and explicit spatial and semantic representations drives the formation of a working model of the information space's shape. During initial exposure to a new information space, users manifest a predictable pattern of inter-action whereby they seek to find key points or landmarks within the space to which they can return. Typically this is the top level or entry point, and users are known to repeatedly land on this point when disoriented or experiencing difficulties (Lee *et al.* 1984; Norman 1991). Interestingly, this pattern emerges on paper as well as in digital information spaces (McKnight *et al.* 1990).

However, while most hypermedia researchers have interpreted such behaviour as a motivation to pursue studies of features such as links, trails and browsers, that is, elements that allude to explicit spatial properties of the interface, our research places as least as much emphasis on the role of semantics. Semantics also appear crucial to the process through which members of a discourse community learn to shape their communications over time. Repeated interactions seem to give rise to regularities in presentation and consumption of discourse (Bazerman 1988). While the term 'genre' has been used to cover such regularities, the definition of genre remains contested though it points to a set of issues about information organisation that would appear crucial to digital spaces (and is beginning to be addressed at last in the literature on electronic document systems, see for example, Reiffel 1999). From the design perspective, we can conceive of communities of users learning over time to identify regularities in form that convey clues as to position in the narrative, stage of argument, likely following sequences and points of closure. Such knowledge could serve as a top-down interpretative framework that interacts with the incoming stimuli from the visual display to form an active model of information space in which the user is currently resident. This conceptualisation seems in line with emerging cognitive models of reading and, in particular, the readiness effect (Gerrig and McKoon 1998), which manifests itself in parallel processing and fast-acting resonance in memory for new input of information that has been previously processed.

The interaction of spatial and semantic processes is represented more easily in a figure (see Figure 7.1). This is a conceptual representation of the processes described in this chapter. The information display is perceived by the user who then creates a dynamic working model of the information space contingent on current contents and their format. Relevant spatial attributes (layout, image placement, length of text, window size, navigation icons, etc.) combine with activated memories of just processed information as well as semantic attributes of the information genre applied top-down (expected form, style, sequencing, meaning, etc.) to create a continuously updated

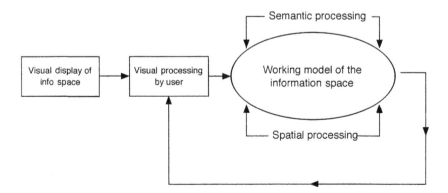

Figure 7.1 The spatial-semantic model of shape perception.

and modifiable dynamic representation of the information space for this interactive task.

Seen this way, different users can apply different semantic processes to the creation and maintenance of their working model of the information space. Experience with a form or a specific space will afford such processes. Lack of experience will lead to a reliance more on spatial properties alone. Errors in performance result from lack of certain details in the user's working model. Analysing the type of errors made could then indicate the reliance any one user is placing on certain cues; for example, the novice's tendency to target title words in the text as indices of location in the whole document is a form of 'matching bias' long known to cognitive psychologists who have studied problem-solving (Johnson-Laird 1983). Shape can thus be operationalised as the structural components of the working model that the user forms of an information space. Such properties will always be implicit in an information space, but part of the interface design problem is to offer improved explicit representations that are compatible with the processing tendencies of humans.

From shapes to digital genres

Clearly, a problem for digital document designers and users is the lack of agreed genre conventions that will support the formation of shapes. One could try to transfer wholesale the existing structures of the paper world but since the earliest days of hypertext research it has been seen that this rarely works (for example, Wright and Lickorish 1988) and tends to diminish the very qualities of paper that work so well in the analogue medium. Furthermore, such copying fails to exploit the potential of digital presentation to be reconfigured according to cognitive compatibility, task dependency or any other criterion we care to invoke. What is required therefore is an understanding of the properties of information that are most important in developing a sense of shape. It is important to understand the extent to which different users exploit, or find exploitable, these properties.

The long-term process of working model formation is complex. The surface-level semantic and visual cues manipulated in several of my own studies are insufficient to explain performance. While the precise form of cue clearly varies across major superstructural sections of this information space, experts are exploiting deep semantics to gain a sense of order. It is this level of processing that novices clearly lack, forcing them to rely solely on surface-level information that they can see or infer.

Where users of an information space possess structural knowledge of the domain, they will expect to apply deep semantic knowledge to the task in hand. Violations of conventional form or genre will detract from usability. This detraction might induce a time cost (less efficiency), an output cost (less effectiveness) or an affective cost (lower satisfaction) with the application. Similarly, total reliance on semantics to convey structure will also incur performance costs since experts also exploit spatial cues. More interesting is the weighting of the effects – perhaps explicit spatial information can lessen the impact of structural violations at the semantic level, or perhaps semantics overwhelm explicit spatial cues. As yet, we have insufficient data to answer this and it is seems to me that information visualisation work has not yet addressed such issues directly (see for example, Card *et al.* 1999).

For the novice, interaction with a new information space must necessarily be driven by spatial information. In this case, interfaces should convey explicitly the linkages, layout and high-level organisation of the space to minimise disorientation. The mainstream hypermedia literature has useful advice here but it is insufficient for all but small and/or infrequently used spaces. All the evidence we have accumulated on the spatial-semantic issue suggests that spatial cues are coupled to semantic information as the user naturally seeks to abstract regularities in the information space. The human cognitive system continually seeks to apply existing knowledge to new information. Even with information for which there is no historical form, it is likely that genre formation starts to occur in the mind of the user almost immediately. Thus, we need to study how spatial aspects interact with semantics as the user seeks to abstract regularity from the new space. Certainly the interaction of semantics and spatial components seems better to explain how the mixed showing of spatial ability correlates with user behaviour in hypermedia environments as highlighted by Chen and Czerwinski (1997).

Beyond gross differences in the knowledge base of users, there may exist deep psychological differences based on ability or preference to deal with semantic or spatial cues or some weighted mix thereof. Such a possibility is not only suggested by the model but is explicitly pointed at by research in education that shows differential effects for hypermedia across learner types (see Dillon and Gabbard 1998 for a review).

Approaching the issue from the perspective of digital document designers, it is intriguing to speculate how quickly digital genres may form. If the spatial-semantic perspective is correct, it is conceivable that a genre can be formed rapidly in any situation where there exists a community of users interacting repeatedly within an information space, and this space manifests regularities that participants can perceive. Our analysis of the rapid emergence of web home pages as a genre (Dillon and Gushrowski 2000) shows that preference for certain home page elements is positively correlated with their frequency of occurrence in pages at large. Thus new genres are continually being formed, and the sociocognitive analysis of information shapes might offer us a cohesive means of unpacking the complexities of this process.

Conclusions

The spatial-semantic model assumes all information spaces convey structural cues to the user that are differentially exploitable depending on the user's knowledge, experience with the genre, and interactive behaviour patterns.

This dynamic combination of spatial and semantic information gives shape to information for the user. While the greatest source of difference between users might be the level of semantic processing they can apply, there may be other individual differences to consider, such as the user's preference for spatial or semantic cues, which might reflect a deep cognitive style difference. This remains a concern for future research.

Designers and evaluators of digital information spaces should be informed by a clearer appreciation of the various spatial and semantic affordances manifest in any context. To this end, user analysis of the kind envisaged by Dillon and Watson (1996) could be employed to gain a more reliable and valid estimate of user differences and requirements.

The shape construct is part of a broader push for a sociocognitive analysis of interaction that seeks to reflect the interaction of multiple levels of information processing in the human, thereby blurring the rather rigid traditional boundaries between physical, perceptual, cognitive and social perspectives that dominate current thinking. It is the present author's contention that we can only fully understand the human response to technology by adopting a multilevelled analysis of interactive phenomena that vertically slices through the time-based layers of standard social science analyses. Only then can we move the field beyond designing for usability to designing for augmentation. In the following chapter a framework is presented which serves as a summary of these issues for the purposes of design.

8 TIME: a framework for the design and evaluation of digital documents

The man who doesn't read good books has no advantage over the man who can't read them

Mark Twain

Theory versus empiricism: the strained role of users in the design process

Human factors research is often criticised for being piecemeal rather than coherent, evaluative rather than predictive and addressing specific issues in a way that leaves little scope for generalisation of findings (Chapanis 1988). This is reflected in the ergonomics practices carried out in systems design where human factors are often considered at a stage too late to effect better designs, that is, the human factors specialist is seen as having a primary role in testing instantiated designs rather than influencing the initial specifications. Yet, given the much sought-after opportunity to become involved in the early specification phase, human factors professionals (be they usability specialists, information architects or interface designers) suffer from a lack of the conceptual tools and techniques necessary to overcome resistance from engineers seeking better inputs than vague or inflexible guidelines and exhortations to 'know the user'.

As outlined in Chapter 2, the standard philosophy underlying much human factors work is that of iterative user-centred design involving the development of prototypes and their subsequent evaluation, leading to further proto-typing and so forth. While such an approach, properly executed, makes the development of usable technology more likely, it is a suboptimal process that can prove extremely expensive in terms of time and resources. Few design companies are therefore willing to invest the necessary effort to iterate through several cycles. This had led to the attempted reduction in the number of iterations needed and a move to bring human factors inputs into the design process earlier (Gould *et al.* 1997; Eason 2001). Current emphasis is on rapid prototyping facilities that allow designers to mock-up disposable simulations quickly and cheaply. These can help but, even then, the quality of the original prototype is dictated by the accuracy of the designer's conceptualisation of the intended users, as outlined in Chapter 2. One can now prototype to high fidelity with less effort than ever before, but this ability has no bearing on one's ability to create the right design first time.

Card *et al.* (1983) argued for an applied science of the user that is theory based rather than empirical, using a common framework to provide insight and integration. By this they meant that analytical techniques that do not require any empirical input could be used at the earliest stages of the process. As noted in Chapter 4, they proposed a constrained version of cognitive psychology, with its emphasis on the information processing aspects of humans as the most suitable vehicle for this science and argued that if it is to have an impact, such a science must be based on task analysis, calculation and approximation, which would lead to quantitative performance models of users. The role of such a framework would be to encapsulate some relevant knowledge of the user (often termed a user model, user typology or user stereotype) and/or the task (similarly termed task model, etc., by some writers) that could provide guidance to the designer specifying the system. According to Card *et al.*, the true role of an applied psychology is to provide such performance models in quantitative form for designers.

In reality, the apparent extremes of frequent empirical iterations and formal theory-led designs are merely opposite ends of a continuum; ends in which few user experience practitioners (as opposed to academics) permanently reside. More common is a mixed approach linking empiricism to theory and vice versa, with a bias towards empiricism due to the perceived lack of relevant (that is, applicable) theoretical models at this time. The mixed approach is probably inevitable as both extremes are impossible to implement absolutely. All observation is theory impregnated according to most contemporary philosophers of science (see for example, Chalmers 1982). Thus any artifact is going to be coloured by assumptions about the user, however implicit, when it is being developed prior to testing (and in philosophical terms one could argue that even the design of the experiment to test a system reflects underlying belief systems and is therefore theory impregnated). The empirical route to design can therefore in no way claim to reject totally theoretical perspectives of the user in favour of experimental facts. However, complete theories of human performance in HCI (or anywhere else for that matter) are nonexistent and any design based on theoretical models alone must be evaluated by empirical means to ascertain its true level of usability.

Therefore, a practical goal for frameworks and models in HCI is to guide the derivation of suitable initial designs which, by virtue of their accuracy or utility (and these terms are not equivalent), reduce the number of iterations required before an ultimately acceptable design is achieved. Evaluations would subsequently act as confirmation or rejection of the design (or parts thereof), and, if the latter, lead to refinement of both the resulting system and the theoretical framework underpinning it. The value of frameworks or models therefore lies in their ability both to reduce iterations and to be modified, if necessary, in the light of data from users and subsequently applied to other designs.

Frameworks and models: a clarification of terminology

The terms framework and model tend to be used interchangeably in the HCI literature, with model being the dominant descriptive term for such theoretical views of users. However, for purposes of clarity, the present work will draw a distinction between them in the manner described previously by Whitefield (1989). He describes a framework as a generic representation of the important aspects of the user and a model as a specific representation of those aspects in relation to a task. In this sense a framework provides the perspective of the user (that is, reader) for all instances of interest while the model is derived according to the interaction of particular task demands and the user. In the present case a framework is proposed of the generic aspects of the reader, which it is hoped will support the derivation of more specific models of reader–text interaction for particular tasks.

Four important criteria impinge on any proposed framework beyond the obvious one of utility.

1   It must be accurate. This is not to say that it must offer a precise picture of the user or reader and document interaction being supported, but what it offers should be correct in the sense that it describes real factors or aspects that influence the reading process.
2   It must be relatively noncomplex. Invoking psychological descriptors or cognitive structures in a form suitable for nonspecialists to use and apply is a difficult but necessary part of a good framework.
3   It must be suitably generic to be of relevance to more than one application. Just as the reading process covers myriad texts and tasks, designers should be able to utilise the framework describing this process for guidance on the design of more than one text system.
4   It should be modifiable. This does not mean that it must be altered every time it is used but that it should be capable of being adjusted in the light of feedback.

The following descriptive framework is an attempt to satisfy all four criteria. The next section outlines the framework in detail.

The TIMEframe

In developing this framework, a representation was sought of the issues that emerged repeatedly in design contexts involving a variety of hypermedia and other electronic documents (for example, software manuals, process handbooks, research project archives, academic journals). The intention of the framework is to provide those developing digital information resources with a simple way to conceptualise the human factors influencing the usability of the created artifact. TIME is based on 10 years of investigations of human information usage from a user experience perspective. This framework was termed TIMS in its original explications (Dillon 1994) but the revised version reflects a slightly modified labelling and reemphasis of the same underlying components.

The TIMEframe is predicated on the following assumptions of human information usage:

1   Humans explore and use information in a goal-directed manner to 'satisfice' the demands of their tasks.
2   Humans form models of the structure of and relationship between information units. Repeated formation and application of such models leads to the formation of schematic forms and, ultimately, document genres.
3   Human information usage consists in part of physical manipulation of information sources.
4   Human reading at the level of word and sentence perception is bounded in part by the established laws of cognitive psychology.
5   Human information usage occurs in contexts that enable the user to apply multiple sources of knowledge to the information task being performed.

In form, TIME is a qualitative framework and is proposed for use as an advanced organiser for design, as a guide for heuristic and expert usability evaluation, and as a means of generating scientific or testable conjectures about the usability of any electronic text. The framework raises for consideration four key factors affecting usability. They are:

- a task (T) that reflects the reader's needs and uses for the material;
- an information model (I) that consists of the user's mental model of the information space;
- the manipulation skills and facilities (M) that support physical use of the material;
- the ergonomic variables (E) influencing the perceptual processing of words and images.

Human information usage is thus conceptualised as a process involving the abstraction of meaning in accordance with the shifting focus of the user, and embedded in an informational context that impacts all four levels of analysis. The framework is applied by walking through a task the user will perform, and articulating the TIME activities as they occur. To make this clearer, it is necessary to explicate the TIME components in more detail.

Tasks (T)

Users interact with information purposively, to obtain data, to be entertained, to learn, etc. To do this they must decide what it is they want to get out of the resources to hand, determine how they will tackle the information space (for example, browse or read start to finish, follow a link or ignore it for now, etc.). Furthermore, during the task they must review their progress and, if necessary, revise aspects of the task.

This notion of intentionality in usage gives rise to the idea of planning, which evidence suggests is relatively gross, taking the form of such intentions as 'locate a reference that is relevant', or 'see what this site is about'. However, plans can be vaguer. Reading an academic article to comprehend the full contents seems to be unspecifiable, the reader is more likely to formulate a plan such as 'read it from the start to the finish, skip any irrelevant or trivial bits, and if it gets too difficult jump on or leave it'. Furthermore, such a plan may be modified as the reading task develops; for example, the reader may decide that she needs to reread a section several times, or may decide that she can comprehend it only by not reading it all. In this sense planning becomes more situated (see for example, Suchman 1988) where the reader's plans are shaped by the context of the ongoing action and are not fully specifiable in advance.

Information modelling (I)

Readers possess (from experience), acquire (while using) and utilise a representation of the document's structure that may be termed a mental model of the text or information space. Such models allow readers to identify likely locations for information within the document, to predict the typical contents of a document, to know the level of detail likely to be found and to appreciate the similarities between documents, etc.

The information model results from the user's attempts to organise the information space's contents into a meaningful structure. Where this model is weak, navigational difficulties will occur as the user cannot call on an accurate representation of the organisation of the information space.

It is important to note that the model is not purely the result of bottom-up processing. Users may have a preexisting model of how the information is organised, and thus what is initially a model becomes, with use, a map of a specific text. Where no model exists in advance, it is hypothesised that a map can be formed directly (though it may require more effort on the part of the user). Similarly, where repeated map formation suggests regularities in the model, then where this is communicated across time and other users, a genre may be formed. In digital library (DL) research, there are many questions as yet unanswered about information models.

Manipulation (M)

Interacting with information involves a substantial amount of physical engagement. With paper, such skills are acquired early in life and are largely transferable from one text form to another. Most readers can use their fingers to keep pages of interest available while searching elsewhere in the document or flicking through pages of text at just the right speed to scan for a particular section, but beyond these actions, manipulation of documents becomes difficult. When one then considers manipulation of multiple documents these limitations are exacerbated.

Large digital documents are awkward to manipulate by means of scrolling or paging alone but 'point and click' facilities and graphical user interface qualities have improved this, at least in terms of speed, although the present author still longs for the digital equivalent of the finger that is less permanent than a bookmark and serves the temporary holding of a place in space. The lack of standards in current electronic information systems means that acquiring the skills to manipulate documents on one system will not necessarily be of any use for manipulating texts on another. Obviously, digital systems afford sophisticated manipulations, such as searching, which can prove particularly useful for certain tasks and render otherwise daunting tasks (such as locating thematically linked quotations from the complete works of Shakespeare) now manageable in minutes rather than days. Yet electronic search facilities are far from a guarantee of accurate performance.

Ultimately, the goal is to design transparent manipulation facilities that free the users' processing capacity for task completion. Slow or awkward manipulations are certain to prove disruptive to the usage process. The typical delay experienced by web browsers at busy times is the most obvious example of this. The framework raises these issues as essential parts of the information usage process and therefore important ones for designers to consider in the development of electronic text.

Visual ergonomics (E)

The final element of the framework reflects the visual ergonomics factors. It is at this level of analysis that we deal with the activities most typically described as 'reading' in the psychological literature or the experimental HCI literature to date. Thus, eye movements, fixations, letter/word recognition and other perceptual, linguistic and (low-level) perceptual and cognitive functions involved in extracting meaning from the textual image are properly located at this level.

In terms of design for digital libraries, this level cannot be overlooked. Ability to effectively read electronic text, for example, is contingent upon image quality and any application that fails to address this basic level will not be redeemable to users in terms of content or task support. Thus, the issues at this level can be seen as necessary (though insufficient) determinants of digital library usability. With the proliferation of screen reading in all sizes, from handheld personal digital assistants (PDAs) to giant home theatre screens, the importance of visual ergonomics is increasing.

So far, the basic components of the framework have been described. A schematic representation of the framework is presented in Figure 8.1. As shown, the elements are all related and collectively framed within the context in which the activity occurs.

Interactions between the elements

When users interact with a document they engage in multiple rapid acts involving the various elements in the TIME framework. For example, a user might start by targeting certain information in a document. To gain the fastest answer it is likely that the user will apply her information model of the document's structure. This would involve $T \rightarrow I$ and $I \rightarrow T$ exchanges, as well as numerable $I \rightarrow M$ and $M \rightarrow E \rightarrow M$ activities as the user scans the text headings and moves through the text. Finally, if successful, the user will locate the relevant section and proceed to read the words in detail, at which point the major human factors issue is one of screen ergonomics as the user reads the text.

As this happens, it is possible for the designer to analyse the task of the user in such a scenario in order to get a sense of the various proportions of the user's efforts that are placed at each level (see a worked example of this in Chapter 9). Thus a task that requires multiple searches and quick

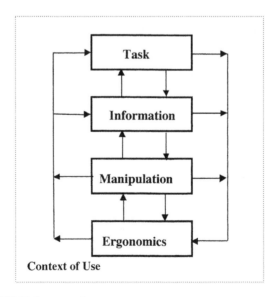

Figure 8.1 The TIME framework.

location of details will place manifest different sequences of T → I → M → E interaction than a task involving once-off location followed by lengthy reading.

According to the framework there are 12 possible interactions between these elements that can occur. These will be described individually.

Task model to information model (T → I) When a task is formulated the reader usually interprets or mediates its formulation and expectation of its outcome in terms of her model of the information space. For example, if the task is 'Find the reference in the text to Jakob Nielsen', the reader applies the model to narrow the search space and produce an inference such as 'It is more likely to be in the section labelled "Related Work" than in "Results"'. This is a routine and rapid occurrence.

Task model to manipulation skills and facilities (T → M) Where an information model does not exist (when this is a reader's first exposure to a text type, for example), the reader, upon formulating a task, may proceed to manipulate the text without any knowledge of its layout or contents. This may be as simple as opening a book (or electronic file) with no better intention than reading it from start to finish until the target information is located. In terms of the framework this is a case of direct interaction between the T and the M elements. It implies that absence of an information model at the outset does not prevent text usage.

Task model to visual ergonomics (T → E) In cases where the text is short and available, for example a single-page memo on one's desktop, even manipulation facilities may be unnecessary. Similarly, during a particular subtask of a larger one, a reader may engage only T and E to perform that subtask, for example locate a word in a paragraph currently on screen or on the open page. This example highlights the fact that it is possible to perform some reading acts without engaging either the I or the M elements.

Information model to task model (I → T) A model of the information one is dealing with can influence the type of task one tries to perform with it and aid accurate specification of that task. For example, if a reader's goal is to find out about particular theories of child development, a model of the text (for example, an introductory psychology textbook) could suggest that the book was inappropriate but it might offer suggestions for further reading. On the other hand, such an interaction between I and T elements could occur where, after reading the text for several minutes, the reader's evolving model might indicate that the text is unlikely to contain the form of information required and therefore the task needs to be respecified.

Information model to manipulation skills and facilities (I → M) The interaction between these two elements in this direction is likely to be of the form: model directing manipulation, for example the information being sought is at the end of the text, therefore page or scroll to the last chapter. Such rapid interactions should characterise phases of many reading situations.

Information model to visual ergonomics (I → E) Again, this is only likely to occur for particular tasks and very short texts. An example might be identifying the sender of a one-page letter. In this case one's model of the letter form suggests that an address may be provided at the top of the page or a signature will be present at the bottom. Once the letter has been opened and the page unfolded further manipulation activities can be bypassed.

Manipulation skills and facilities to visual ergonomics (M → E) Once the task and model aspects have been applied and the text is of the type that will require manipulation, an interaction between the M and E elements occurs. A simple example is the reader turning a page to allow reading to commence.

Manipulation skills and facilities to information model (M → I) Such an interaction might be expected to result when, faced with an unfamiliar text, the reader manipulates it and induces the formation of a primitive information model (the 'flick through to see what's in it' approach).

Manipulation facilities to task model (M → T) In this instance the information flows directly from the element concerned with text manipulation to the task model. Though rare, an example might be when a reader finds that she cannot search for a term and therefore cannot perform the task as originally envisaged.

Visual ergonomics to manipulation skills and facilities (E → M) Where reading is interrupted by a page break or screen end an interaction with manipulation facilities occurs to facilitate further ergonomic activity with the text. This is logically distinct from the M to E activity described earlier, which refers to activity occurring prior to E activity.

Visual ergonomics to information model (E → I) Visual ergonomic concerns may interact directly with the information model by providing information suitable for the formation of a model to the reader, thereby supporting the formation of a map; for example, noting the occurrence of a particular word or phrase as a potential landmark in the document.

Visual ergonomics to task model (E → T)   The direct interaction between the two extreme elements in the model may occur in this direction when, for example, information is read which solves the immediate task or sub-task; for example, if, when searching for a word or phrase in a certain section, the reader perceives it and thereby resolves their immediate task without requiring further manipulation or model activities.

In practice, it is not expected that such neat interactions occur in isolated units. Meaningful engagement with a document is more likely to result in multiple rapid interactions between these various elements. For example, a scenario can be envisaged where, reading an academic article for compre-hension, task model concerns interact with the information model to identify the best plan for achieving completion. This could involve several T → I and I → T interactions before deciding perhaps to serially read the text from start to finish. If this plan is accepted then manipulation facilities come into play and serial reading commences. The M → E interaction and E → M interaction may occur iteratively (with occasional E → I interactions as distinguishing features are noted) until the last page is reached, at which point attention passes back ultimately to T as the reader considers what to do next.

The speed and the iterative nature of the interaction between these elements is also likely to be such that it is difficult to demonstrate empirically the direction of the information flow. In many instances it would be about impossible to prove that information went from M to I rather than the other way, and so forth. However, this does not preclude examination of these elements and their interactions in an attempt to understand better the process of reading from a human factors perspective. The elements reflect the major components of reading that emerged as important from the studies earlier in this book and are intended as a broad representation of what ergonomically occurs during the reading process.

Why a multilevelled framework?

TIME is multilevelled in that it brings together concerns with human response at all levels – from the physical to the sociocultural. To use a DL environ-ment requires engagement at the physical and perceptual levels to manipulate and view information. Cognitive processes are involved in comprehending and navigating the information space. Social factors influence the forma-tion of information models and the contexts in which DLs are employed. To model information requires users not only to acquire the form of representation generally evaluated as navigation knowledge, they must also develop an appreciation of the content of the information space in semantic terms. To do this, users seem to learn to respond to conventions in information space that they notice with repeated use, and which experienced members of a discourse community utilise to convey their ideas. The term shape seems to more adequately convey this set of representational cues, and, as such, invokes social determinants of cognitive representation that cannot be conveniently labelled as cognitive or social, but are best understood as sociocognitive aspects of interface design (Dillon and Vaughan 1997 for a detailed explication of the shape construct).

Similarly, the task component of the model is not intended to represent user task in a limited idealised form akin to the classic engineering approaches of Card *et al.* (1983) or Drury *et al.* (1988) but to allow for description of the range of intentions users are likely to bring to a digital library application as they seek to satisfice information needs. In this way it is advocating scenario identification and an awareness of the necessary updating and shifting intentions of the user embedded in a context of information usage (see for example, Rosson and Carroll 2001).

TIME as design tool

The TIME framework is intended to be used as a means of understanding interaction in context. It seeks to avoid prescriptive guidelines for design and instead offer designers and implementers a systematic framework within which to think about the interfaces they are providing to their resources.

In the first instance the framework is useful as a guiding principle or type of advance organiser of information that gives the designer an orientation towards design enabling the application of relevant knowledge to bear on the problem. Thus when faced with a design problem requiring digital documents, for example, the designer could conceptualize the intended users in a TIME framework and thereby guide or inform their prototyping activities.

Second, by parsing the issues into elements, the framework facilitates identification of the important ones to address. This framework suggests four major issues to consider in any information context: the user's task and their perception of it; the information model they possess, must acquire or is provided by the artifact; the manipulation facilities they require or are provided; and the actual 'eye-on-document' aspects involved. All are ultimately important, though depending on the task and the size of the information space, some may be more important than others (for example, for very short texts of one screen or less, manipulation facilities and information models might be less important than visual ergonomics, although task contingencies must be considered here before such a conclusion could be reached).

In the third instance the framework provides a means for ensuring that all issues relevant to the design of electronic text are considered. It is not enough that research or analysis is carried out on text navigation and developers ignore image quality or input devices (and vice versa). A good digital library interface will address all issues.

The above applications consider the uses at the first stages of system development. However, the framework also has relevance to later stages of the design process such as evaluation. In such a situation the framework user could assess a system in terms of the four elements and identify potential weaknesses in a design. This would be a typical use for expert evaluation, perhaps the most common evaluation technique in HCI. This form of use has been most commonly made of the framework by the present author and colleagues.

Certainly an immediate application is to enhance our critical awareness of the claims and findings in the literature on HCI. For example, it is still not untypical to hear statements to the effect that 'the web is better than paper' or 'reading from paper is faster than reading from screens'. TIME suggests caution in interpreting such statements. If human information usage involves all elements in the framework then the only worthwhile statements are those that include these. Thus, the statement 'hypertext is better than paper' is, according to the TIME perspective, virtually useless since it fails to mention crucial aspects such as the context of use, tasks for which it is better, the nature of the information models required by readers to make it better or worse, the manipulation facilities involved, or the image quality of the screen that affects the standard reading processes of the reader. We know all these variables are crucial since we have over 15 years of research investigating them and trying to untangle their varying effects.

Thus, TIME suggests that statements about interaction with digital documents, to be of value, need to be complete and make explicit each of the elements in its claim. Thus, a more useful statement would be: 'For reading lengthy texts for comprehension, for which readers have a well-developed information model, on a scrolling window, mouse-based system, screens of more than 40 lines are better than those of 20 lines, though both are worse than good quality paper'.

Notice that the truth content of the statement is not what is at issue here. It is the completion of the statement that is important. Incomplete statements (ones not making reference to all elements of the TIME framework) are too vague to test since there are unlimited sources of variance that could exist under the general headings 'hypermedia', 'paper', or 'is better than'. A complete statement might be wrong, but at least it should be immediately testable or (if you are really lucky and informed) refutable, given the existing body of empirical data the field has amassed.

Coupling these activities to the operationalisation of usability is another possible application of the framework. It is difficult to meaningfully derive criteria for effectiveness, efficiency and satisfaction for users of an application without some informed view of the tasks they will be performing and the experience and skills they possess. Stepping through the elements of the TIME framework, particularly the task and information model components, focuses attention on key issues that affect these criteria-setting operations. In this way, TIME can be the basis for a document usage task analysis.

Qualitative versus quantitative representations

The framework presented here is a relatively simple representation of those issues found to be of importance to the usability of an electronic text. They are described in this framework qualitatively, that is, their actions are presented in general terms rather than being specified formally. The absence of rules of exclusion/inclusion or numerical values necessitates greater interpretation by an eventual user of this framework than a quantitative framework or model of HCI or reading. This is intentional, a matter of choice (and some necessity given current knowledge) rather than a failure on the author's part to specify further the framework's components. In the present section the case for such a framework is presented by considering it in the light of the usage of typical quantitative models more commonly expounded in this domain.

As outlined earlier in the book, traditional psychological models of the reading process are often very detailed, postulate the existence of numerous cognitive structures and processes, and tend to concentrate on isolated or limited aspects of the reading process such as word recognition, sentence processing or eye movements. It has been argued at length in Chapter 4 that the level of detail provided by such models of human information processing are too low to be applied in HCI and that, in the case of reading, this severely hampers the development of usable electronic text systems. The form of modelling common to cognitive psychology is mirrored sharply in the attempts of human factors professionals to describe or model human behaviour at the computer interface. As detailed in Chapter 4, the major research effort has concentrated on developing formal models of a quantitative kind for designers to apply at the specification stage of product design.

Advocates of the quantitative approach cite precision, nonambiguity of terminology and ability to calculate design trade-offs as major advantages of such models (see for example, Harrison and Thimbleby 1990). While this may be true for models such as GOMS or cognitive complexity theory (CCT) when used for very specific analyses (and there is little by way of confirmatory evidence of this yet), there are two underlying assumptions in this view which are directly pertinent to the present work. One is that designers find such quantifiable outputs relevant and the other is that the human performance and behaviour one is interested in can be reduced to suitable numeric functions. There are many further criticisms of these formal methods that could be made which are not directly pertinent to this discussion, for example, What of their accuracy? Why do independent users of the methods often derive different models of the same problem? or Why are they so difficult to use?

According to Landauer (1987) such models do not tell a designer how to design a good system in the first place (which is what designers really want to know). Instead, they just advance the moment when evaluation can first be carried out to the pre-prototype stage, that is they are a measurement tool rather than a creative design aid.

The present framework takes such shortcomings as its starting point and is intended to offer a conceptual aid to electronic text design that addresses such problems. First, a designer does not require sophisticated knowledge of human cognition or the psychology of reading to comprehend the framework. Obviously, detailed psychological work underlies concepts such as information models, task models, etc., but the designer can consider the basic issues without possessing such knowledge. The more knowledge of human cognition that a designer possesses, the more critically and usefully she may be able to apply this work, but such knowledge is not a prerequisite for use. Second, unlike most formal methods, the present framework does not require the use of a formal language or sequences of rules to support interpretations of likely user behaviour. It is intended only to draw attention to issues such as image quality and information organisation in the first instance so that the designer realises what is important in a design, not to provide a means of calculating design trade-offs in terms of performance times. Third, the framework covers the full range of behaviour described as reading as it impacts on system design, not just a particular subset of it. It is intended to cover reading as it pertains both to proofreading and scanning of lengthy texts, for example, or to using textbooks or magazines. Finally, it is suggested that a successful electronic text system is one that addresses all four elements of the framework in its design, therefore a designer can employ it to guide her initial specifications; that is, it is a design aid more than a measurement tool.

A suspicion exists that qualitative approaches are inherently vague, are more likely to be rejected by engineers (who supposedly prefer numbers to words) and, in the light of the aggressive 'selling' in the human factors literature of the 'hard' quantitative approaches, are somehow less 'scientific', however that is measured. This need not be so, however. The literature on design (not specifically HCI-related), outlined briefly in Chapter 2, has clearly demonstrated that designers tend to rely heavily on heuristics, intuition and 'try it and see' approaches rather than the rigid hypothetico-deductive logic-based approaches manifest in trained scientists. Qualitative models could well offer the form of guidance more suited to this type of problem solving than some time-consuming but powerful quantitative approach.

As stated earlier, implicit qualitative models abound. All designers and ergonomists, in fact everyone involved in the development of a product, from the marketing department to the specification writers, have an implicit model of the users and tasks the end product will support. These models just vary in detail and accuracy depending on its possessor's role. The sales representative presumably has a view of the user as a customer while the marketing person might view users as belonging to certain job, skill and economic categories. These views or representations of the target users are 'models' as such.

For present purposes, the main interest is in the models possessed by the designers and ergonomists. The latter participant, by virtue of her probable training in a human science such as psychology, is likely to model the user as an information processor with cognitive dispositions, skills, habits and preferences. This model will probably include detailed knowledge of cognitive components such as short-term memory, long-term memory, mental models, etc. and their potential impact on the usability of a computer system. On the basis of task analysis and previous experience, skilled ergonomists can derive a set of user characteristics for input to design specifications. In a very real sense then, this is a form of qualitative modelling. By extension, quantitative model proposers must have their own qualitative models on which they base their formalisms.

Designers, as described earlier, will always implicitly model the user in drawing up specifications. However, their models of the user and task tend to be ill-formed and vague, often based more on intuition than on facts. Yet the resultant usability of a product is largely determined by the quality of the implicit model underlying the design and mistakes made at this point in the product life cycle are considered to be the most expensive to rectify. A progressive aim of human factors inputs, therefore, must be to improve this model, either by quantitative or qualitative means. The evidence, on balance, would suggest that a suitable qualitative model is likely to be more relevant to designers than existing formal quantitative ones.

What needs to be improved is the explication of these models. Vague descriptions of user characteristics are probably better than nothing but guidelines and handbooks of design principles are rarely successful. Opting to present a more structured view in terms of a framework describing the relevant components of the user–system interaction, as embodied in the present framework, is likely to have more relevance.

Rasmussen (1986) advocates the use of qualitative models in this sense. He argues that quantitative modelling concentrates on one level of behaviour, particularly sensorimotor in well-practiced tasks, which is inappropriate for the type of higher level cognitive functioning of interest to many designers. For him, the major distinction between the two forms of model is not that one is respectable or scientific and the other intrinsically soft and vague, but that the qualitative types concentrate on broad categories of behaviour while the quantitative focus on specifics. He rejects the traditional argument that the former are merely undeveloped or premature quantitative models and states that designers of computer systems might well find qualitative models of direct relevance to their work in the design of any system where users have some choice on how they will work.

This is an important but often overlooked difference between current and past technologies. Interactive computer systems afford greater user control than traditional mechanically engineered machines that had to be operated in a set manner. In the case of electronic text design it has been strenuously argued that the quantitative approach is not appropriate. Whether this is a function of current knowledge limitations or inherent failings in the approach is not of direct concern in this book, but philosophically at least, the present author's inclinations are in the latter direction. The type of knowledge needed by the designers I have worked with in two continents over the past 15 years is often of the generic qualitative kind. Furthermore, the act of reading as it is interpreted in the present book involves behaviours and cognitions too broad to fit the 10-second boundary of the classic model human processor approach of cognitive psychology. Therefore, regardless of the ultimate success of a quantitative analysis of cognition, qualitative models do seem to have relevance worth pursuing as design aids in HCI.

The utility of the proposed framework

What use is this framework to designers of electronic text systems? I propose three potential uses:

1   In the first instance the framework is useful as a guiding principle or type of advance organiser of information (Ausubel 1968) that gives the designer an orientation towards design, enabling him to bring relevant knowledge to bear on the problem.
2   Second, by parsing the issues into elements, the framework facilitates identification of the important ones to address. This framework suggests four levels of issue to consider in any context: the user's task and their perception of it; the information model they possess or must acquire; the manipulation facilities they require for a given task or scenario of use; and the actual 'eye-on-text' ergonomics involved.
3   In the third instance the framework provides a means for ensuring that all issues relevant to the design of electronic text are considered. It is not enough that research is carried out on text navigation and developers ignore image quality or input devices (and vice versa). A good electronic text system will address all issues (indeed it is almost a definition of a good electronic text that it does so).

The above applications consider the uses directly to designers at the first stages of system development. In this sense the term designer encompasses any user experience or human factors professionals seeking to influence the specification of an application. However, the framework also has relevance to later stages of the design process such as evaluation. In such a situation the framework user could assess a system in terms of the four elements and identify potential weaknesses in a design. This would be a typical use for expert evaluation, perhaps the most common evaluation technique in HCI.

Outside of the specific life cycle of a product, the framework has potential uses by human factors researchers (or professionals less interested in specific design problems) in that it could be used as a basis for studying reader behaviour and performance. The framework is intended to be a synopsis of the relevant issues in the reading process as identified in the earlier studies. Therefore, it should offer ergonomists or psychologists interested in reader–system interaction a means of interpreting the ever-expanding literature in a reader-relevant light. These issues are discussed further in the next chapter.

9   Applying the TIME framework

Cross even a stone bridge only after you have tested it

Korean proverb

Introduction

The framework as described in the previous chapter derives from the various analyses on readers' classifications of texts, descriptions of their usage and numerous experimental investigations of their models for information spaces. In this chapter I show how TIME contains the elements which capture typical reader–document interactions and can be used to predict user response in both experimental and usability walkthrough contexts.

Validity: observing TIME in users' verbal protocols

This study examined readers' performance in extracting answers to questions from a short text on the subject of winemaking presented in four different designs, one paper and three electronic. The envisaged scenario was one where an individual, armed with the document, staffed an enquiry service where people would ask relatively straightforward questions on such topics as the largest winemaking regions of France or the meaning of terms such as 'second fermentation' and so forth. No previous knowledge of wine was required as the answers to all the questions were available in the text.

For present purposes the analysis focuses on the concurrent verbal protocols elicited from readers while also examining some of the performance data. The aim is to determine if the elements in the TIMEframe capture the essential cognitive processes of the user during an information task of this nature.

Readers, tasks and texts

Sixteen experienced users participated in the study, nine male and seven female, age range 21–36. The text under consideration was *Introduction to*

Wines by Buie and Hassell (1982), a basic guide to the history, production and appreciation of wine. For the test we employed two hypertext versions (TIES and HyperCard), an MS Word version (all presented on screen) and a printout of the Word file for the paper condition.

The paper format consisted of 13 A4 pages of text with no figures and would thus be most aptly described as a booklet or essay-type text. In order to place a structure on the document that would facilitate its presentation as a paper text, the topics were retained in the linked groups of the hypertext original but ordered from start to finish in a manner that seemed intuitively sensible to the experimenters. Thus an introduction was followed by a general overview of the processes involved in manufacturing wine before specific countries and regions were presented. This structure was retained faithfully for the word processor version.[1]

Readers were required to use the text to answer a set of 12 questions. These were specially developed to ensure that a range of information retrieval strategies were employed to answer them and that the questions did not unduly favour any one medium. The answers to all questions were specifically mentioned in the text. A four-condition, independent subjects design was employed with presentation format as the independent variable. The dependent variables were speed, accuracy, access strategy, readers' estimate of document size and verbal protocols.

Readers were tested individually in the usability laboratory. Any questions the reader had were answered before a 3-minute familiarisation period commenced, during which the readers were encouraged to browse through the text. After 3 minutes the readers were asked several questions pertaining to estimated document size and range of contents viewed. They were then given the question set and asked to attempt all questions in the presented order. Readers were encouraged to verbalise their thoughts and a small tie-pin microphone was used to record their comments. Movement through the text was captured by video camera.

The application of the framework to the location task

The TIME framework suggests that there are four major components to the reading task in any context, each of which is represented by a rectangular box in Figure 9.1. This schematic representation provides a descriptive model

---

1 In order to test this intuitive arrangement for suitability a quick pilot test was carried out by the author. This involved asking readers to order a set of cards, each of which had a term such as 'Bordeaux', 'Production', or 'Aperitifs', rather like a list of contents. Three readers were each asked to group these into what they perceived to be a suitable single-document structure. The results confirmed the structure of the experimenters; that is, groups were formed out of countries and subordinate regions, wine manufacture, and particular wines and grapes.

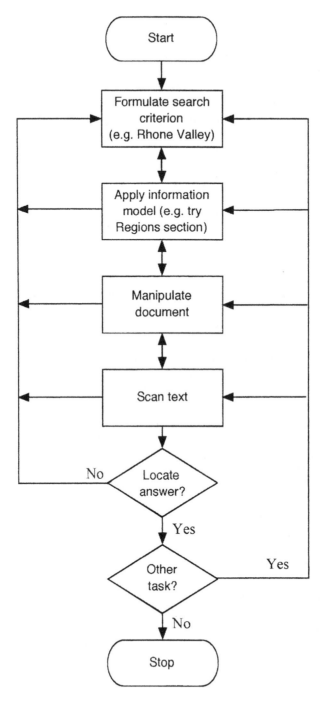

Figure 9.1 A schematic model of readers' behaviour on an information location task.

of the likely sequence of events involved in performing the experimental tasks with this document.

Initially, the reader must formulate a means of resolving the task. It is hypothesised that for the type of tasks involved in the present experiment, the reader will identify a search criterion from the question and attempt to obtain an answer by finding a relevant match to that criterion in the body of the text. For example, if the question is 'What type of wines are produced in the Loire region?' it is assumed that the reader is most likely to select 'Loire' as a target and locate references to this in the text until a pertinent section on wine types is located. However, selection of any other target is possible and allowable within the framework.

Once a satisfactory search criterion is identified, TIME suggests the reader's attention switches to her information model, which is used to guide the search for a matching term. It has been shown that the information model of certain text types is well-formed and supports such applications, but in the present context it is expected that the uniqueness of the text would be unlikely to afford a detailed model, particularly at the outset. However, even for unique texts, exposure to them facilitates the development of a map and it is likely that even though a reader lacks an information model at the start, after several tasks she will begin to acquire one. This should be apparent from the verbal protocols.

The reader could, for certain tasks and applications, bypass or overcome any inherent limitations in her information model by employing the search facilities of the computer. If, at the task processing stage, the reader deduces a specific and infrequently occurring term, the search facilities available on two of the four versions could be used to locate the required text directly. Certainly, for the tasks that can be resolved in this way, advantages should be conveyed to readers using suitable electronic versions (in this case HyperCard and the word processor version). This should be apparent from their performance data.

For other tasks, the manipulations should be less straightforward. For example, one of the tasks required readers to compare two sections of information before gaining the information necessary to provide an answer. One could imagine the readers using a paper document opening the first relevant section and keeping a finger on it while searching for the second, prior to flipping between them to obtain the answer. This is a typical reader–paper text interaction but a difficult one to mimic without (and sometimes even with) windowing facilities on screen. Neither of the hypertext applications in this experiment supported windowing of this nature, although this was possible with the word processor version. In such tasks, one would hypothesise advantages to paper over electronic text. The advantages and disadvantages of manipulating text should also be manifest in the readers' protocols.

Once the reader has searched and manipulated the text, the framework suggests that a scanning type of reading follows. In other words, it is not expected that readers will read large amounts of text in a serial fashion while performing these tasks but will jump and skim read sections looking for cues and target words. From the work on proofreading text on paper and screen reviewed earlier, it is clear that in general, the advantages lie with paper. Although few researchers have examined scanning as opposed to proofreading, there is no reason to assume that a similar advantage to paper does not hold for this type of reading style too.

If the target is successfully located at this point then the task is completed and the reader can start on the next one. This initiates a sequence of events similar to those just outlined, though with each subsequent completion it is expected that the information model becomes more elaborate (that is, a mental map of the document is being formed) and familiarity with the requisite manipulation skills grows. This should also be apparent in the protocols.

Hypotheses

As the model of user activity deduced from the framework suggests, a single prediction about the most suitable application for these tasks was not possible. Paper would seem to have certain advantages over all electronic versions in some circumstances (phases of E activity) while it is possible to see advantages for certain electronic versions in others (for rapid M activities). However, two general hypotheses suggested themselves on the basis of the framework:

1   The size of the document, and the lack of any specific model of the information space that users have on first approaching the text, should make information modelling easier in paper initially. This should manifest itself in problems estimating the document size and greater navigational difficulties in the hypertext conditions.
2   The verbal protocols will reveal user cognitions that map directly and sufficiently to the elements of the framework.

Results[2]

Estimating document size

Hypertext users had difficulty assessing the document size accurately while readers in the linear conditions were far more accurate. After familiarisation with the text, readers were asked to estimate the size of the document in pages or screens. The linear formats contained 13 pages, the HyperCard version

2   Some of the data presented here are used by kind permission of my coworkers at the time: Cliff McKnight and John Richardson of the HUSAT Research Institute in the UK.

contained 53 cards, and the TIES version contained 78 screens. Therefore, raw scores were converted to percentages. The responses are presented in Table 9.1 (where a correct response is 100, scores above and below this number reflect over- and underestimates, respectively).

Readers in the linear format conditions estimated the size of the document reasonably accurately. However, readers who read the hypertexts were less accurate, several of them overestimating the size by a very high margin.

Navigation

As stated in Chapter 3, a general measure of navigation is nonexistent but relies on the interpretation and operationalisation of the concept by individual researchers. For present purposes it was assessed by examining the proportion of time spent viewing the Contents/Index (where applicable) by each reader as a percentage of total time. This provided a highly objective behavioural measure rather than any indication of subjective difficulty. These scores are presented in Table 9.2.

Table 9.1 Readers' estimates of document size

Subject	Condition			
	TIES	Paper	HyperCard	Word processor
1	641.03	76.92	150.94	92.31
2	58.97	92.31	56.60	76.92
3	51.28	76.92	465.17	100.00
4	153.84	153.85	75.47	93.21
Mean	226.28	100.00	187.05	90.61
SD	280.41	36.63	189.84	9.75

Table 9.2 Time spent viewing Contents/Index as a percentage of total time

Subject	Condition			
	TIES	Paper	HyperCard	Word processor
1	53.28	2.72	47.16	6.34
2	25.36	1.49	19.10	13.93
3	49.50	10.24	17.50	12.87
4	30.84	5.36	23.40	7.54
Mean	39.74	4.95	26.79	10.17
SD	13.72	3.88	13.81	3.79

Very large differences were observed between both hypertext formats and the linear formats. A one-way ANOVA revealed a significant effect for condition ($F_{[3,12]}$ = 9.95, $p < 0.005$). Even using a more rigorous basis for rejection of the null hypothesis in post hoc tests than the 5 per cent level, that is, the $10/k(k-1)$ level, where k is the number of groups, suggested by Ferguson (1959), which results in a critical rejection level of $p < 0.0083$ in this instance, post hoc tests revealed significant differences between paper and TIES (t = 4.90, d.f. = 6, $p < 0.003$), between word processor and TIES (t = 4.16, d.f. = 6, $p < 0.006$) and between HyperCard and paper (t = 3.06, d.f. = 6, $p < 0.03$). Thus, interacting with a hypertext document seems to necessitate heavy usage of browsers or indices in order to navigate effectively through the information space.

Searching for precise information

It is expected that when readers seek information for which they can formulate accurate search terms, applications that offer such facilities should lead to faster and/or more accurate task completion rates. However, use of the search facilities rests on several factors: the signalling of their presence; the user's willingness to employ them; and the user's ability to use them correctly.

In this study, three of the tasks were supported by the search facilities in the HyperCard and WORD conditions. The mean time per reader on these tasks is shown in Table 9.3.

A one-way ANOVA comparing those applications with search facilities and those without showed a significant effect in the hypothesised direction ($F_{[1,15]}$ = 6.10, $p < 0.03$). Interestingly, not all readers used the search facilities for all possible tasks, as suggested above.

Table 9.3 Mean times to perform tasks supported by search facilities

| Subject | No search facilities | | Search facilities | |
	TIES	Paper	HyperCard	Word processor
1	194.33	66.00	112.00	97.00
2	79.67	34.67	44.67	59.33
3	281.67	233.00	39.33	81.33
4	122.67	170.67	36.33	76.33
Mean	169.59	126.08	58.08	78.50
SD	88.43	91.99	36.11	15.57

Evidence for the interactive elements from readers' protocols

The data described above demonstrate that the framework can be employed to guide reasonably accurate predictions about reader task performance. However, the main aim of the validity study was to examine the extent to which the framework can be seen as a fair representation of the issues involved in reader–text interaction. To test for this, the verbal protocols of the readers were examined.

Each protocol was transcribed from a videotape according to a predefined classification scheme derived by the present author in conjunction with the other members of the research team. This scheme captured the verbal utterances, the time they occurred, the actions performed by the readers and any further behaviours deemed relevant by the evaluator such as readers having difficulties with an application or making an error in their answer. The author subsequently examined these in order to identify verbalisations that mapped onto the framework. Protocol data are rich and complex and therefore not easily reduced to a simple presentable form. In the present context isolated sections from a selection of readers in a manner akin to Suchman (1988) seem to be the most appropriate means of highlighting the existence of the elements of the reading process described in the framework. Accordingly, several examples from different readers are presented to provide an insight into the reading process from the point of view of the reader.

Box 9.1 describes a typical section of protocol from one reader. The protocol presentation format involves a timescale in seconds, a transcription of the user's verbal protocol and a description of user action on the system.

Clock	Comment	Action
00.00	OK, I'm going to the Index to see if any of these wines are mentioned	Selects Index button
00.05	Don't appear to . . .	Reads Index
00.25		Selects Contents button
00.35	I'll have a look in . . . probably the Making of Wine section . . .	Reading Contents
00.58	I'll go to Sweetness because it's the only term in the Contents list that really . . . refers to taste	Selects Sweetness

Box 9.1

This represents a short section (1 minute) of a reader looking for specific information in a hypertext version. While on the surface it might seem trivial, it is obvious that much activity is occurring in the reader's mind. First, without hesitation, he has accessed the Index. This is logical behaviour as indices provide references to and the locations of material contained in the text. However, this implies that not only does the reader have expectations of what he will find in that section of the text but that he has formulated a means of task resolution (or at least a first step towards it) and acted upon it rapidly. Without some form of awareness of the order of the text, the manner in which it might be structured and the access mechanisms available within it, that is, knowledge of the information space and the manipulation facilities available, such rapid and meaningful processing would not be possible.

It would be almost impossible to determine the order in which the cognitive processing went for such an interaction. Obviously he knew his task and had a formative map of the document (at this stage the reader had already spent 3 minutes familiarising himself with the text prior to commencing the trial). These elements must have interacted, but whether that interaction was of the form $T \rightarrow I$ or $I \rightarrow T$ is probably not important from a designer's point of view. In reality it was probably a case of cyclical interaction between both. What is certain is that the reader decided on a course of action, manipulated the information space and rejected the first attempt at problem resolution in a matter of seconds, exactly the type of task flow captured by the TIME framework.

In fact, in the first few seconds of this example it is possible to see all four elements of the framework in action. Once he decides to go to the Index, the Manipulation element operates (button selection) and the Reading Processor takes over as he scans down the list of topics to reveal no mention of the target item.

The first interaction of another reader, this time in the MS Word document condition, is presented below.

Clock	Comment	Action
00.10	Right, I'm going to go to the Contents and try and get some idea of where this might be	Drags scroll bar up Reads Contents
00.24	I think it's probably in the Making of Wine	Still reading Contents
00.38	Oh is it just Sweetness and Body?	Referring to items in content list
		continued

Clock	Comment	Action
00.42	I'll just go there and check if they're the two . . .	Scrolls down the text then clicks in the scroll bar until she reaches relevant section. Reads text
00.58	So the two things are Sweetness and Body	Finds answers

Box 9.2

Again the interaction of several subprocesses of the reading task can be identified. Upon reading the question, the reader's first decision is to go to a section of the text that might offer guidance pertinent to the task. As before, the expectation that Contents sections adequately map the information available in the text, and the fact that they are located in a particular place, should be noted. At this point an interaction between T and I has occurred followed by a quick burst of M and E activity to move up the text on screen and identify the relevant section as the desired one.

Upon reading the Contents (mainly E activity) this reader identifies two subsection headings that seem relevant to her representation of the task (T activity) and decides to go to that section (I followed by M activity) to check if they are relevant. At this point she scrolls down the page before deciding that the scroll bar would be a better option (M activity interacting with T activity and I activity – the reader must decide at some point that the information is located suitably 'far away' to justify a faster or more efficient scrolling method). Upon arriving at the relevant section (rapid E, I and M activity) she ceases scrolling and starts to read the text (E activity) until she confirms her opinions.

At this stage in both examples only the crudest information model of the text exists. As this is a unique text type to which the readers would not have been previously exposed there would be no existing structural models for them to employ and only maps could be formed. With increased exposure this map is elaborated allowing more accurate predictions of what is located where and the type of information available within it. Thus one observes numerous comments to the effect that 'I've seen this before somewhere' or 'I've an idea where this one is' and so forth. The following example is from a reader using the word processor document who is now on his fifth task.

Here we see the emergence of landmark knowledge of the information space ('I've seen nothing on that' . . . 'I've just passed a section on . . .' etc.) and knowledge that can allow the reader to make informed judgments about locations that enable him to evaluate (and in this case, reject) hypothetical locations of information ('Grapes on page 1? . . . No'). Such processing is only feasible where the individual has at least a rudimentary map of where things are in the document, what they are next to, whether or not he has seen them before and what type of information a section may contain. As suggested by the framework, this type of knowledge is likely to be picked up by readers as they become familiar with a document.

Also pertinent here are the limitations of the manipulation facilities for scrolling text in word processors. In going to the Index in the first instance this user drags the scroll bar down to the bottom of the window with the result that he overshoots the start of the index section and needs to scroll gently back up (a total activity taking approximately 10 seconds). This is similar to, but probably more awkward than, using indices at the back of books, that is, overshooting to the back of the book so that you need to page back to the index section is a common experience during reading but shouldn't waste 10 seconds. Technology should make such a process simpler (or unnecessary), not more difficult. This is an action that is well supported in the 'point-and-click' facilities common to the digital systems.

Clock	Comment	Action
06.07	to the index then . . . I haven't seen anything on this before	Drags scroll bar down
06.09		Scrolls to top of Index
06.17	Grapes on page 1? . . . No	Reading Index
06.25		Scrolls up to page 2
06.30		Scrolls up to top of section and reads
06.42		Scrolls down to top of next section
06.56		Scrolls back and rereads this section
07.00	It must be in the body of the report then . . .	
07.30		Spends 30 seconds scrolling down and reading the following four subsections
		continued

Clock	Comment	Action
07.33	OK, dessert wines	Locates answer
07.42		Reads next question
07.45	I've just passed a section on aging . . .	Refers to new question as he scrolls up to that section

Box 9.3

These examples are typical of the protocols elicited in this study. All protocols were examined in this way and it would be overwhelming (not to mention unnecessary for present purposes) to reproduce them all at this level; however, a complete protocol for one reader is presented in the appendices to give an indication of the process. The framework proposed here seems to be supported by the evidence from protocols of readers using both paper and hypertext versions of a document in the following ways:

1   There is evidence of the existence of each of the elements, that is, readers verbalise thoughts which confirm attention changing from attributes of task, information model, manipulation and serial reading of the text.
2   There is evidence of the interaction of the elements in both linear and nonlinear fashions (that is, interactions do not necessarily follow the strictly linear sequence $T \to I \to M \to E$ but combine in sequences which reflect the reactive nature of the reading process).
3   There appeared to be no evidence from any of the protocols to suggest other elements need to be incorporated in the framework, all relevant verbalisations and behaviours proved classifiable as belonging to one of these categories (though of course, these categories or elements are rather general and hide many complex cognitive issues as stated earlier).[3]

The elements in this framework provide a parsimonious account of the types of utterances and behaviours solicited from readers performing routine tasks with a text. No attempt has been made here to compare the verbal protocols between conditions to identify any differences that might exist between them. For example, it is possible that readers in the hypertext condition manifested fewer comments on their information model than

3   Not surprisingly, there were verbalisations that were deemed irrelevant, such as comments to the experimenter regarding the time, the ease or difficulty of particular tasks or quips about the study. However, these amounted to a very small proportion of the total data elicited and can safely be considered inconsequential with regard to the framework.

readers in the paper condition, or most comments in all conditions were about manipulation rather than scanning issues. This is a question for further research.[4]

In summary, the framework posits the existence of four elements of concern to the reader of any text. The verbal protocols support the existence of these and suggest that the interactions between these elements is of the general form described in Chapter 8. The framework provides an adequate account of the type of processes carried out by readers of both electronic and paper texts.

Utility: using TIME to predict user response to e-documents

Overview

The present experiment examines the use of TIME to derive predictions about user performance with a digital document. The test document is an electronic academic journal. On the basis of the evidence reviewed in this book, paper articles would often be preferred and prove more usable. However, for the other forms of use to which journal articles are put electronic text is likely to offer certain benefits. This text type would therefore seem a useful test of the utility of the framework in determining the types of activity for which each form is strongest.

Applying the framework to the description of academic article usage

A journal reader approaches a text with a task or set of tasks that she hopes to resolve, no matter how ill-specified. According to the framework it is suggested that the readers apply their model of the text structure to the task in order to direct their activities. Thus they decide if they need to look at some part of the article more than the other, where that part is located, where the other relevant articles might be and so forth. If the electronic version maintains the paper structure there should be no differences between the

---

4 A further reason for not pursuing a more quantitative analysis of the protocol data is that such a procedure is difficult to perform objectively. In the examples described above one can identify the general sequence of activities and their relationship to the framework. However, to force every protocol into an all-or-nothing categorisation whereby each utterance is classified as belonging to one or other element in the framework would hardly be informative and would lead to a mass of numeric data that added little or nothing to the present description of the framework. Many of the interesting utterances concern the rapid interaction of several elements or the continual interchange between two nonadjacent elements. If the quantitative categorisation was to take account of all elements, the possible interactions between elements and the timeline associated with each utterance and try to relate these to each of the four conditions in the experiment it is likely that the resultant analysis would overcomplicate the data to the point of meaninglessness.

media at this stage, that is, their well-developed model would be equally relevant to either medium.

Readers then manipulate the text and locate the section(s) relevant to their needs. Traditionally this would have been difficult with electronic text but the availability of hypertext applications eases the manipulation task considerably, particularly where text is broken or 'noded' into selectable chunks. In the case of the article, jumping to various sections and headings should be facilitated on screen, though location of particular blocks of text within a larger body of text is unlikely to be so easy.

Once at the relevant section it is probable that readers adopt one of two reading styles: straight serial reading from the start or quick scanning. In reality, readers probably adopt a mixture of both, switching as needed *in situ*. Where the reader adopts a serial reading style, paper is likely to be better than hypertext. This seems probable given the weight of evidence showing a performance deficit for proofreading speed from screens and the difficulties readers have with lengthy electronic texts. However, the differences between the two media are likely to be lessened where the amount of text to be read is small (for argument's sake let us say a screenful). Where the reader is scanning the material and it is not lengthy, there is likely to be little difference between the media (assuming image quality of the screen is good). If the text is broken into various small subsections within a section and the reader has an idea where she wants to go, hypertext should convey advantages over paper.

Accordingly, a simple descriptive model of user behaviour for a particular task, for example checking a detail in the method section such as the number of readers employed or the type of equipment used, could be derived. In circumstances where the paper article's structure is retained in hypertext we would assume no difference between the media until the M and E (serial reading processor) elements of the framework are invoked. At this point one would expect an advantage to the hypertext version for getting to a headed section, but an advantage to the paper version for the scanning or serial reading phases of performance. A descriptive model of such a task is represented in Figure 9.2.

According to this model the task basically consists of quick target identification, a rapid application of the information model (I), and then a sequence of M ↔ E interactions. The latter interactions dominate the task according to the framework though each element is not used in equal proportion. Given the task involves scanning text in a specified (that is, given) area, there is likely to be a bias towards phases of E activity. Manipulation (M activity) should be rapid while serial reading (classic E activity) could be extended.

From what is known about reading from contemporary screens it is clear that for serial reading from screens (E activity), paper should be approximately 20 per cent faster. While manipulation can prove problematic for electronic texts, the 'point and click' approach of hypertext where the

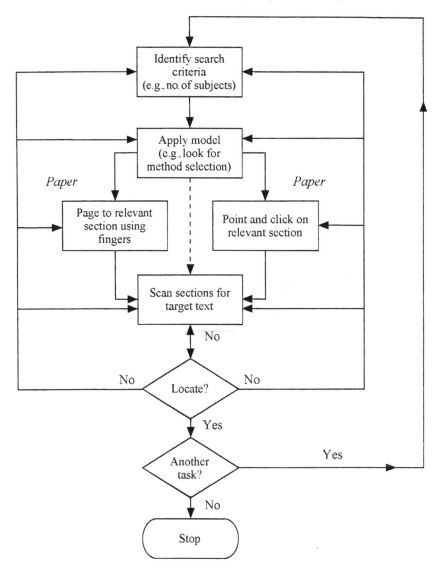

Figure 9.2 A schematic model of readers' behaviour for journal task.

targets can be directly addressed from one screen should provide advantages to the digital form. However, given the relative proportions of time estimated to be spent in either activity (that is, the majority of the task is E activity, not M) this would not be enough to offset the reading speed advantage to paper.

Although there should be an overall advantage to paper it is possible that for targets not requiring large phases of E activity the ease of manipulation with hypertext might prove sufficient to give it an advantage. This would occur in situations where the target sentence was situated at the start of an opened section. Conversely, the speed advantage to paper should maximise the differences between the media for targets situated towards the end of lengthy sections. These differences suggested by the model are tested in this study.

Readers, texts and tasks

Twelve readers (age range 22–35, mean age 27; six male/six female) participated voluntarily in this study. All were professional researchers experienced in the use of academic journals and computers.

Two academic articles were selected according to similarity in terms of length, broad subject matter and conformance to the general superstructure of articles described in Chapter 7. Top quality photocopies were made for one condition and hypertext versions for the other, with links from all major headings (enabling complete reduction to four headings at one extreme and complete exposure of full text at the other).

Readers were required to locate 32 sentences in the academic articles. These were divided into four task blocks of eight sentences each so that each reader located sentences in both texts using both media. The sentences were presented on cards that stated in which section of the text (Introduction, Method, Results, Discussion) the sentence was to be found. The task was designed to be a simulation of the situation common to readers of these texts, which is checking a detail of the paper when they have a fairly reliable notion of where the target sentence is located but are not likely or able to use the search facilities (that is, they cannot formulate an appropriate search string).

The sentences were selected so that equal numbers came from each section, they were of approximately similar length (that is, less than two printed lines) and did not contain eye-catching words (for example, all-capitals) or symbols (for example, numbers). The electronic versions of the texts were made so that line lengths were comparable to the paper ones thereby ensuring a similar typographical form for both media. As well as being situated in particular sections, the target sentences were further distinguishable in terms of their within-section location. Thus in all but the Method section for which it would be impossible, the sentences in each section were selected so that they were at the start or towards the end of the relevant section. The qualification for being located at the start was determined by the presence of the sentence in the first full screen of text that was presented upon opening a section in the hypertext. This allowed further analysis of reader's performance, that is, the effect of scanning large amounts of text on either medium.

A two-condition (paper × screen) repeated measures design was employed using two texts. All readers performed the task twice on paper (once per text) and twice with hypertext (once per text) with order of texts and presentation medium counterbalanced to avoid any systematic ordering effects. The independent variable was presentation medium and the dependent variable was speed of task performance.

Readers performed the experiment in a room with the computer placed on a desk which was free of other materials allowing them to perform both the paper and the hypertext tasks without changing desks. The investigator sat at the edge of the desk in a position where he was able to see the screen and the document being read at all times.

All readers were introduced to the specific workings of the digital version employed here. All were then encouraged to interact until they were comfortable with it, at which point they performed five trial tasks to consolidate their training. If they still experienced any difficulties further blocks of trial tasks were available. However, no reader asked for or appeared to require extra training. Readers were informed that they would be timed for each individual task.

Hypotheses

Given the task and the conditions under which they are presented it was expected that there would be only two levels of difference between the media: the manipulation and the skim reading ones (M and E elements) as outlined above. Given the variation in locations and the estimated proportion of time spent on each activity, three experimental hypotheses were proposed:

1   There would be a significant completion rate difference between the two media with paper proving faster than hypertext (due to the presumed majority of E activity).
2   Readers should locate information in lengthy sections of the texts faster with paper than with hypertext (joint M and E activities, with the latter proving decisive).
3   Readers should locate information in short sections of text faster with hypertext than with paper (joint M and E activities with the former proving decisive).

Results

The effects of medium, text and question on performance

A three-way, $2 \times 2 \times 8$ ANOVA (medium by text by question) with repeated measures on all factors was carried out on the data. Although the texts were selected for similarity it was decided to test for text in case it was producing an effect that was not dependent on the variables controlled for in their selection such as idiosyncratic writing style or vocabulary. The output from this analysis package is summarised in Table 9.4.

Table 9.4 ANOVA summary table for utility experiment

Source	d.f.	SSQ	MSQ	f	p
e-text/paper (A)	1	33189.85	33189.85	14.909	0.0029
Text 1/2 (B)	1	2470.51	2470.51	1.7418	0.2119
AB	1	133.01	133.01	0.2187	0.6524
Question (C)	7	122309.50	17472.79	12.655	0.0001
AC	7	29399.67	4199.95	2.7579	0.0131
BC	7	16546.74	2363.82	1.3393	0.2429
ABC	7	20704.99	2957.85	1.43105	0.2045
Error	352	565404.37	1601.71		
Total	383	790158.64			

These results indicate a significant effect for medium, question and the interaction between medium and question. There was no significant three-way interaction effect or significant effect for text type as expected.

Clearly, there was a significant effect for medium with paper proving to be faster than hypertext for this set of tasks. Thus null hypothesis 1 can be rejected. Mean time per task with hypertext was 52 seconds compared to 33.5 for paper (that is, paper was approximately 35 per cent faster than hypertext, which is slightly larger than the range of speed differences between these media typically reported in the literature).

The significant effect for question was also expected. Searching for targets in text sections of varying length should lead to speed differences with shorter sections affording faster location. This requires no further explanation. Mean times for each location confirmed the direction of the differences; for example, location times for questions 3 and 4 (Method section) were 14.65 and 16.46 seconds, respectively, while mean location times for questions 1 and 2 (Introduction section) were 40.77 and 56.60 seconds, respectively.

The effect of target position on performance

According to the framework, readers should have been able to locate sentences that occurred in the short sections of the texts (for example, in the Procedure subsection of the Method section) faster in hypertext than on paper. The opposite should hold for sentences embedded in longer text sections such as the discussion (though where sentences occur at the start of such sections hypertext should regain the advantage). Thus the advantages of either medium should be tested between questions.

The significant effect for question and, more importantly, the significant interaction effect for medium and question give an indication of what happened in the present case. By examining the mean times per task for each medium it is clear that the advantage to paper is most obvious for tasks involving the scanning of large sections of text. No such difference holds when the target sentence is located in shorter sections. The unweighted marginal means for task by medium are presented in Table 9.5. p values were obtained by using the marginal means to calculate a value for t according to the formula $t = $ (mean1 − mean2)$/\surd$(mse/n1 + mse/n2) (as described by Ferguson 1959, p. 238), where mse = 1522.86 and df = 77 (that is, $1 \times 7 \times 11$).

These data provide a better view of the results than the overall difference between media. Far from being enormously better than hypertext for this type of task it can now be seen that the advantage to paper is maximised mainly for location of material that is situated towards the end of lengthy sections. Although this difference was nonsignificant for target sentences in the Introduction, it was still large and in the hypothesised direction. This supports experimental hypothesis 2. For targets that occurred in the first few paragraphs of a section or in sections that did not contain large expanses of straight text (for example, the Method sections), readers performed as well with electronic text as with paper. However, they did not perform significantly better, as predicted.

The significant effect for targets in the early parts of the Results section runs counter to the suggestion drawn from the framework. Examining these tasks on the recordings confirmed what had been suspected by experimenter observation, that is, readers regularly missed the target on first exposure to the section and read serially through to the end before returning to the relevant part of the text. In all, eight readers failed initially at least once to locate the target even when it was first on screen. Of these, six failed to do so on more than one occasion and two readers missed the same target on two occasions (that is, they reread a section more than twice before locating the target). Although there were no equivalent data records of readers in the paper conditions, the experimenter noted only one reader doing this.

Table 9.5 Mean times (seconds) per target for each medium

Target	Hypertext	Paper	p
Introduction/early	46	35.5	ns
Introduction/late	63	50.5	ns
Method 1	15	14	ns
Method 2	17	16	ns
Results/early	75	39	0.05
Results/late	93	42	0.05
Discussion/early	50	46	ns
Discussion/late	57	24	0.05

Discussion of results

The present investigation was intended to simulate the type of task performed by readers when searching for a specific piece of information in a familiar text type. It was hypothesised that this would be the type of reading task for which hypertext would, in part, offer suitable, perhaps even advantageous, support.

The task consisted primarily of two components of the reading process, manipulation and skimming of texts, which were related to two elements of the framework. Hypertext offered clear speed advantages for getting to a section. This supports the original view of hypertext as a potentially advantageous medium for manipulating large texts.

However, once at a section, readers performed faster with paper, particularly when the target sentence was not immediately obvious. This is explicable, at least in part, in terms of the image quality hypothesis. The statistically significant advantage to paper overall predictably emerged as a result of concatenating all the actions into one performance score: total time. Since the proportional time spent scanning was normally greater than that spent manipulating the text, this weighted the final measure in favour of the task component known to be best supported by paper (in other words, the task lessened the influence of one of digital text's major advantages).

These results support two out of three of the hypotheses derived from the framework. However, the predicted advantage to hypertext for locating text in short sections of text never materialised. In fact, the general trend of the results has been to distort the predicted differences further in the direction favouring paper than was expected, that is, the reading speed difference was larger than normally observed.

In terms of the descriptive model derived from the framework there are obviously few, if any, modifications required to explain these findings. It led to the accurate prediction that paper would be better both overall and for locating sentences situated in the later parts of lengthy sections. The shortcoming of the other hypothesis seems to result not from shortcomings of the framework or any of its postulated elements but from the task and sensitivity of the measures employed.

The sentence location task was intended to simulate the type of reading scenario where a person searches relatively familiar material to identify a certain detail. Behaviourally, at least, the experimental task employed here matched this. However, cognitively it is difficult to maintain the comparison. Readers in the present situation reported trying to locate the required sentence in a simple pattern match fashion; that is, they focused on a key word or phrase in the target sentence and searched the text closely for anything that matched this without considering the meaning of the material being attended to. In the real reading situation one would expect the reader to be more influenced by the context of the material and seek to relate the content of the currently attended-to paragraph to his information needs.

In this way, real reading would involve the narrowing of the search space according to the context of the target's location; for example, if a reader wants to find a sentence about difficulties with the experimental procedure he is likely to appreciate the relevance of other words or sentences which mention other problems or procedural issues. There was no evidence of readers in the present study trying to relate the content of the target sentence to the experimental text beyond the cues they were given for searching in a specific section. Indeed, the typical reading style manifested by readers was a straight serial read from start to target (or finish) of the prompted section, scanning every intervening word. In reality, one would expect to see readers jumping about within sections, ignoring paragraphs which the first sentence indicated were unlikely to contain the required details. This would have the effect of speeding up this process and thereby lessening the proportion of task time spent at the E level of the framework with commensurate benefit for hypertext users.

A further source of potential bias in the study was the assumed lack of training required by readers in the digital condition. Although all readers expressed confidence after the familiarisation period with the use of this application for manipulating the text and there were no instances of readers being unable to open or close sections of the hypertext during task performance, most reported afterwards that they would require a lot more use of e-journals in this manner before feeling as comfortable with them as they were with the paper texts. Several readers certainly manifested nonoptimum use during task performance; for example, opening irrelevant subsections and failing to close them before opening another which resulted in large text sections to scroll through if they went back to the 'start' to reread a section.

The design of the experiment, with its demand that readers complete all tasks, failed to allow for any potential speed/accuracy trade-off. As many of the readers overlooked the target on initial exposure and were forced to reread sections, sometimes more than once, their performance (that is, speed) scores deteriorated rapidly. Had readers been given the option of giving up, the speed differences might not have been so large. Furthermore, for most electronic texts, when faced with a situation where a target is proving evasive, the search facilities are likely to be used. These were not supported in the present task as it was an attempt to simulate the type of interaction where the reader has only an ill-formed idea of the specific details of the search, therefore being more likely to use recognition over recall to aid location. Future work would do well to consider such task effects and select accordingly.

In summary, the data generally support the task predictions derived from the TIME framework and serve as an example of how a TIMEframe analysis of a reading scenario can yield relatively accurate insights into user response.

TIMEframes for usability walkthroughs

While TIME can support experimental analysis, the most direct use of TIME is in structuring the walkthrough of a usability inspection. Few designers or evaluators conduct formal experimental trials (regrettably) and while the preceding examples show the validity and utility of TIME as a framework, it is to be expected that a more typical use of any such framework is its use in guiding usability evaluators.

As outlined in Chapter 2, usability inspections are a common expert-based form of evaluation. In a typical scenario, the evaluator is knowledgeable of HCI and user interface design principles, and is required to evaluate the usability of a design for a target user and task domain. In common inspection methods the evaluator imagines how a target user would react to the design and the evaluator walks through a typical task trying to place herself in the mind of the intended user. In so doing, the evaluator often applies design principles or heuristics to help structure the process. There remain many concerns about which guidelines or heuristics are best, but that concern is to my mind less important than ensuring the evaluator can envisage correctly the usability issues as seen by a user. After all, flagging potential heuristic violations is of limited use if these are of minimal interest to a user, and many inspections based on heuristics tend to overestimate the importance of such violations.

The TIME framework provides a structure by raising four levels of concern for an evaluator within any given context of use. So the first task is to identify the context since, by definition, usability is highly sensitive to shifts in tasks, environments and users. This is standard for any usability evaluation and there is nothing unique to TIME in this. Once the context has been adequately described then the use of TIME is as follows.

1 Specify task sequence
   The easiest way to do this is to determine the steps a user must take to complete the task in a satisfactory manner. This should be a task path that follows the actions a user who can complete the task would follow. Number these required task steps from 1 to *n*. For the evaluation walkthrough, treat each step as if the user has arrived there and has the task goal in mind.

2 For each step, estimate the user's cognitive, perceptual and physical actions in terms of:

   • their information model
   • the available manipulation facilities
   • the visual ergonomics of the design.

3 Add comments or raise questions
   At each step the evaluator should note or question any aspects of the design which may fail to meet expectations or which require the evaluator to make assumptions about the user.

4 Produce output

The TIME approach lends itself to producing a schematic, termed a TIMEframe, which outlines the steps, issues and questions.

The most difficult step here is number two, which requires the evaluator to make informed estimates of user cognition and likely behaviour when faced with the specific design at each step. As with all such usability inspection methods, the quality of the evaluation rests more on the evaluator than the method. However, with this method, it is my view that even inexperienced evaluators can produce better estimates of usability than they otherwise would working without its structure. That said, the onus is placed squarely on the shoulders of evaluators to learn as much as they can about the important issues involved in usability and design if they are to be of value in their work. The procedure is summarised in Table 9.6.

Table 9.6 How to conduct a TIMEframe evaluation

Component	*Activity*	*User requirement*	*Evaluator comment*
Task	• Enter steps to be taken from 1 to *n*	In this column, specify what the user must or should know, should remember (from other steps, etc.) or any other assumptions you are making about your users	Here add in any comments, questions or issues for consideration by the design team; for example, design problems, inconsistent features, checks about assumptions, etc.
Info model	• For each step consider the user's model and the navigation cues		
Manipulation	• For each step consider the clicks and scrolls necessary to perform them		
Ergonomics	• For each step, analyse image quality, screen layout and aesthetics		
Context of use	Specify the likely or actual environment, users, tasks and tools being considered for the above activities		

Worked example: a TIMEframe for a digital newspaper scenario

In the following example, the evaluator wishes to apply TIME to the evaluation of a prototype digital newspaper. One simple scenario she is testing is the case of a user looking for a sports score in the body of the 'paper'. I will concentrate on that to explain the process but evaluators should note that one would normally have to repeat this process for several tasks to gain better coverage of a real application.

To conduct the walkthrough an evaluator determines the steps to be taken to complete this task. She determines that an ideal path involves the user taking four task steps (labelled here T–T4), each of which has a commensurate I, M, and E activity potentially attached to it. The walkthrough proceeds with the usability evaluator taking each of these steps and determining, through the lens of the TIME elements, how a user would likely react.

Assuming a relatively experienced user of newspapers, the evaluator considers that a likely behaviour when faced with the task of locating a sports score would involve the application of a well-formed model. It is reasonable to assume that a user would expect to find the 'paper' grouped into sections, much as one does in the world of physical newspapers. From here, the task is assumed to follow a simple sequence of category selection as the user proceeds to follow a likely path to goal. How accurately this set of assumptions matches reality is beyond the TIMEframe itself and needs to be calibrated with reality by the evaluator continually.

The process of conducting such a walkthrough is relatively simple and quick. Each task, once determined as the basis for the evaluation, should be analysable in minutes, assuming the evaluator has some knowledge of design and user interface principles. The information can be simply tabulated and augmented with comments as to user requirements and usability issues noted by the evaluator, to form a TIMEframe, as in Table 9.7.

This output represents a clear summary of steps and the evaluators' concerns. The comments field contains element-specific issues and reactions, or suggestions for further work. Another usability evaluator can examine this output and add or react to these issues, or get a sense of the first evaluator's thinking. For example, another evaluator might think that a user would not have a well-formed model of a digital newspaper and might fail to apply (or even possess) a model from the paper world. In such a scenario, the assumptions about how a user will proceed may be different. The output of the TIMEframe provides a brief audit trail of reasoning that can be shared across the design team or across evaluators to support discussion.

This use of TIME is currently the most common and it is clear to me, having taught it to evaluators both in academia and in industry, that its use in this manner is easy to grasp and that most evaluators can employ it quickly as an aid to their usability inspections. The precise form of use the TIMEframe receives is determined largely by these users, I do not provide a strict procedure, only a set of steps to follow. In this way, I have seen much

Table 9.7 A TIMEframe for the expert evaluation of a prototype digital newspaper

Task steps	Information	Manipulation	Ergonomics	Evaluator comments
1 Find Match result	User must recognise appropriate link	Move mouse to correct option and select	Read screen to locate links	Will user recognise the links? Does the display conform to the user's model
2 Select Sports section	Repeat	Repeat	Must scan list, perhaps consider colour coding?	
3 Select Latest Scores	Repeat	Repeat	OK	Is this step necessary – could we combine latest scores with top section?
4 Find score	Recognise score	Scroll text	Need to improve scannability	Font and layout needs to be optimal here. Allow scrolling or add more selections?
Final points	Info space maps very closely to existing paper forms	Too many clicks to get to a 'hot' area?		Redesign? Minor restructuring only

variability in output. I do not consider this to be a major problem, given the aims of the framework in these contexts and the known variability that exists in all usability inspection methods at this time. It is in that spirit that I offer it here.

Conclusion

The framework was examined here in terms of its potential as a usability inspection method, its adequacy for describing user cognition and its ability to predict user performance in specific instances. Its validity has been tested by parsing verbal protocols of readers into convenient chunks and relating them to the elements in the framework where they have been shown to map adequately and sufficiently. For the purposes of providing a noncomplex representation of reader psychology relevant to text use it seems valid.

The utility of the framework was tested by examining the accuracy of predictions derived from the framework. Over all studies, four experimental hypotheses were derived and tested, three of which were fully supported. No more need be said of these except that they were strong, that is, unidirectional, hypotheses. For the hypothesis that was not supported it is possible that experimental design factors are sufficient to explain the data. There appears to be no need to alter the framework to explain the findings. Thus as a means of predicting the likely performance of readers with electronic texts, the framework has demonstrated utility.

The framework has been used subsequently in a variety of other electronic document design projects and is taught as an inspection method in both academic and industrial settings. In the final chapter, therefore, attention turns to the envisaged future applications of this framework and the lessons that have been learned in this work for the development of future digital document systems.

10 Conclusions and prospects

Where is the knowledge we have lost in information?

T. S. Eliot, *The Rock*, 1934

Introduction

This book examines and describes the reading process in a manner that is intended to support sensible analysis of the user experience with digital documents. The framework outlined in Chapter 8 provides a means of raising awareness of human factors in information usage to designers of electronic text systems. In this final chapter, it is time to reflect on where we are and speculate a little on where we are going in the never-ending drive towards e-books and digital libraries.

Describing reading at an appropriate level of abstraction

The process of reading has been subjected to continued examination by scientists from a variety of disciplines for over a century now. For all that effort, the process has still to be adequately described, and probably never will be, by any one discipline. Psychology has formed our understanding of the cognitive activities involved, while information science has concentrated on the more pragmatic issues of providing people with access to stored material through search engines and associated information architectures. Educators, typographers, librarians, sociologists and others have all applied their discipline's tools and theoretical perspectives, and while collectively progress can be said to have been made, few researchers of reading would claim to have all the answers.

The impact of advanced information technology on the reading process is clearly an issue worthy of examination. With the impetus provided by electronic text in general, and the web in particular, this issue is the focus of much attention and speculation. Current research on reading electronic text is both piecemeal and of little direct use to those responsible for designing these tools, primarily as a result of the unidisciplinary definitions of reading adopted by researchers, the limited tasks examined and the resultant failure of any descriptive framework or model to provide a means of conceptualising the range of issues involved when people read a document.

The variance that exists in terms of texts and reading tasks is of crucial importance in describing the reading process. In order to operationalise these factors in a reader-relevant form, people's perceptions of texts and their characteristic manner of using them have been described. Outside of where and when they are read, all texts are describable by readers in terms of three criteria: why we read them, how we read them, and what general type of information we believe documents contain.

The utility of these straightforward criteria lies in their ability to distinguish between texts according to usage factors and thereby group material in a form that directly supports examination of the potential role of information technology in their use. This sets the classification criteria apart from any other text typology, the majority of which have attempted to classify material in terms of linguistic structures from which mappings to electronic text design are difficult to make. Furthermore, they reflect the view that texts exist in the reader's world and provide affordances to which the experienced reader becomes attuned.

When a text has been conceptualised in these terms one has a basic orientation from which to proceed in further describing the reading process. Thus, given any text, the three criteria can be used to elicit detailed information from readers on the type of tasks it is used for, their manner of interacting with it and the context of typical use. This can be done directly by a researcher or designer based on common-sense reasoning (at some cost) or, as shown, more objectively through structured interviews and the situated simulation method described in Chapter 6.

In this way, the reading process is initially conceptualised in terms of the text and task involved, hence the initial element of the descriptive framework: the task. This immediately distinguishes it radically from cognitive psychological analyses of reading which in many ways can be seen as text and real-world task independent. It also distinguishes the description from the type of conceptualisation offered in information science which is concerned with the range of texts as categorisable entities but offers little insight into how individual readers actually construe or use them once they have been located.

The reliance on task analysis as a function of text classification promotes a level of description that can be seen as predominantly psychological in its concepts yet is atypical (in its breadth) of traditional psychological descriptions of reading. In relative terms it is a higher level of description than that provided by cognitive psychology but a lower level of description than that typically provided by information science. This is obvious from two other components of the descriptive framework: the information model and the manipulation element.

The concept of an information model is well established in the psychological and linguistic literature but tends to be used only as a theoretical construct in discussions on reading comprehension (see for example, van Dijk 1980 or Garnham 2001). The link between this work and the more traditional research on reading is only infrequently made. However, the interviews with readers carried out in the course of this work confirm that the concept is inextricably interwoven with text usage, providing a reader with the means of grasping the organisation of material as well as supporting accurate prediction of the location of material in a text. The experimental work on structure and perception of location lends support to these views.

The manipulation element is perhaps the least likely component of the descriptive framework. Few people, when discussing reading, ever consider the issue of document manipulation to be of central (if any) importance. However, from the literature on reading from screen reviewed in Chapter 3, it is clear that manipulation issues are crucial to the analysis of electronic text. Much reading involves manipulation; from pamphlets to ledgers, letters to novels and manuals to encyclopaedia, document usage invariably requires the reader to open and turn pages, keep fingers in the text portion of interest while opening other sections and so forth. In fact, it is such an inextricable part of the process that without the ability to manipulate material easily, much reading would not be possible (or, at best, would prove difficult) with current print media. The framework emphasises the importance of these activities by including a manipulation element in its structure.

The lowest level of the framework represents behaviour more usually equated with the activity of reading. The visual ergonomics element is the component that covers the process of extracting the message from the text, that is it refers to the contact or interaction between eye and print, so to speak. When an individual examines the text at the word or sentence level, the type of activities common to traditional psychological models of reading such as eye movement, word recognition, lexical processing and so forth are presumed to occur. From the point of view of the framework, these issues are pertinent, but only after, or in conjunction with, the range of behaviours and processes described in the other elements.

The framework therefore describes reading as a task-driven activity involving the setting of goals, the evolution and application of an information model, the manipulation of a document and the visual processing of text images. This is in contrast to the definition of reading as the visual and cognitive processing of textual images typical of psychological textbooks or as the acquisition and usage of documentation, to put it in information science terms. It does not suggest that these are the only issues that can be validly described as reading, nor does it imply that any one of these is more/less important in the whole process. Furthermore, it does not suggest that traditional research paradigms on reading are wrong. Its intention is to provide a level of discourse appropriate to the examination of reading in the context of information technology.

The scope of the framework

Each of the elements in the framework raises an issue or set of issues to be dealt with in the design of electronic text. Thus the reading task must first be understood in the terms of the text type and its context of use. The information model element focuses attention on the reader's representation of the document's structure. The manipulation element highlights the importance of such facilities while the serial reading element raises the issues associated with visual ergonomics. Issues that do not map onto one or other of these elements are, on the basis of this framework, of secondary importance to the design of electronic texts.

This latter point is worth elaborating. No scope for the *explicit* analysis of the reading outcome is provided by this framework. So, for example, the concept of comprehension, among others, is not represented in the framework; yet comprehension is, for many theorists, a crucial component of reading. This is not a return to the theoretical debate on the appropriateness or otherwise of comprehension in the discussion of reading with a statement of the present author's recommendation to exclude it. Rather it is a reflection of the goal of the descriptive framework: to support the accurate examination of human factors issues in electronic text design.

Outcomes, such as comprehension, are considered here to the extent only that they are influenced by the reader's ability to use the text. If technology is designed appropriately, users will be able to gain access to well-presented information in an efficient and easy manner. At this point, it is not clear what more can be done with the technology to ensure the reader makes sensible usage of this material; that is, achieves her goal, finds her reference, comprehends the text and so forth. This choice of outcome exclusion emphasises the general paucity of applicable knowledge available from work on comprehension and similar concepts even for designers of paper texts, which implies that attempting to design electronic text that ensures greater comprehension of material, for example, is not an immediately attainable goal (despite the claims of the technocrats). Obviously, as cognitive science progresses, such goals might become more feasible. They are certainly desirable. However, the present author's view is that, currently, they are not practicable, in the sense that a design process cannot be specified sufficiently to ensure their attainment.

Until we have sufficient knowledge about the relationship between information presentation and subsequent learning or comprehension, then the efforts of electronic text designers concerned with usability should be concentrated on providing the tools to access and manipulate relevant material in a suitable manner. This is not defeatist or pessimistic, however, and is certainly not equivalent to Clark's (1983) position that media cannot be considered a determining factor in learning outcome. The attainment of comprehension or other outcomes are likely to be contingent upon such successful and easy access provided by well-designed systems, that is usable systems are likely to result in greater (or at least faster) comprehension than badly designed ones. A badly organised website, for example, will violate information model expectations and induce a navigation cost that will impact outcome. In this sense the consideration of such issues is not dismissed but is placed in perspective. The information modelling element of the framework, with its emphasis on shape and structure, certainly points directly to issues of comprehension in design that must be considered. The reading process as described in this framework is surely a prerequisite to any desirable outcomes such as comprehension. The present framework's exclusion of such concepts from its description of immediately relevant issues is not a dismissal of them but a recognition of their complexity. Put simply, one would not expect a specification of a product to state that it must be built so as to ensure greater comprehension. Although this might be desirable or even required, the specification would state it in more concrete terms; for example, the system must be faster, more accurate, etc. – manipulable variables which are presumed to reflect, lead to, or correlate with, comprehension.

Interestingly, not all cognitive scientists even consider comprehension to be an issue worth addressing. For example, van Dijk and Kintsch (1983) state:

> there is no unitary process 'comprehension' that could be measured once and for all if we could but find the right test. Comprehension is a common sense term which dissolves upon closer analysis into many different sub-processes. Thus we need to construct separate measurement instruments for macroprocesses, knowledge integration, coherence, parsing . . . Comprehension is just a convenient term for the aggregation of these processes: it is not to be reified, not to be tested for.
>
> (p. 260)

Ergonomically designed electronic media may well impact differentially on some of these processes and thereby affect the learning process. The crucial point for educators is to know which ones and how.

The framework as context provider for research findings

The framework can also be seen as an aid to understanding the human factors literature on reading. As described in Chapter 3, this literature is replete with empirical studies on issues such as the effect of image polarity, scrolling versus paging, large versus small screens and so forth. Interpretation of the various findings can prove problematic and there are contradictions in findings that cannot be resolved without reference to contextual factors.

The framework offers such a context within which to assess the findings of any one experiment. Thus, when Gould *et al.* (1987b) claim that reading from screen can be as fast as reading from paper, the framework supports an interpretation of this statement that allows an informed (that is, nonliteral) acceptance. In this case the framework highlights the fact that electronic text can be as fast for proofreading short texts on an ergonomically optimised screen. However, this does not mean that no speed deficits occur for other tasks or texts even with such optimised screens, where manipulation effects must be considered and information modelling issues emerge as the text space grows. The simple heuristic therefore is: for any statement about the advantages or disadvantages of electronic text, consider its reflection of each of the four elements in the framework. If it misses one (that is, fails to include assessment of each element) then it is incomplete.

The issue of statement completion is worth elaborating. To make a complete statement about electronic text reference must be made to the task, the text, the reader's model, the type of manipulation facilities available and the visual ergonomics. For example, the statement:

'paper is better for proofreading tasks than electronic text'

is more complete than:

'paper is better than hypertext'

but less complete than the statement:

'for proofreading a familiar text form, of 10 pages, paper is better than electronic text.'

But all of these are less complete than the statement:

'for proofreading a familiar 10-page text form, on a typical laptop screen with scrolling facilities, paper is better than electronic text.'

In each of these cases the references to particular components of the framework are easily seen, the differences between them lying in the number of elements explicitly articulated. However, despite a statement's completion, its truth content is another factor. A statement may be complete in the sense implied here, but be wrong. However, this is a separate issue. A complete statement is open to evaluation, in terms of either current knowledge or empirical investigation. An incomplete one cannot be so easily tested. For example, the final statement above is easier to comment on appropriately or to test empirically as valid than the first statement. It befalls researchers and designers alike, therefore, when making claims about electronic text, to do so in as complete a fashion as possible. Similarly, incomplete statements (for example, 'hypertext is superior to paper' or vice versa) must be dismissed as nonsense.

Designing in TIME

The delivery format of a qualitative framework rather than any alternative such as a set of guidelines or a quantitative model was adopted for a variety of reasons. Predominantly, the emerging perspective on reading was not amenable to reliable quantification. There are few aspects of reading and information design that are amenable to such analysis and the reception of other quantitative models of HCI (for example, the GOMS model of text editing) by the design community at large hardly inspires confidence in their applicability.[1]

Guidelines were not adopted as there are several problems with them that are well documented, not least their inherent contradictions and over-generalisations. As Hammond *et al.* (1987) noted early in the history of HCI research:

> If behaviour results from an interplay of factors, so will the ease of use of an interface. These interdependencies are hard, or even impossible, to capture in simple statements. A guideline which is true in one context may well be misleading in another . . . the more complex the interface, the less plausible it is that guidelines will help.
>
> (p. 41)

The qualitative framework is seen as a suitable alternative to both the standard models or guidelines approaches. It represents a stylistically simple way of presenting a set of complex ideas and supports 'unpacking' of the components to facilitate more detailed analysis. By representing reading as the interaction of a small number of elements it focuses attention on the range of issues to be considered and their possible interrelationships.

The term 'unpacking' is meant to imply that other forms of advice could be derived from a framework such as this. Guidelines, for example, could be 'unpacked' from particular components. For example, 'when transferring paper to the web, retain the useful structural components of the original' could be a guideline derived from the IM element, or 'for presenting text on screen ensure image quality is high' could be similarly derived from the visual ergonomics element. I have not attempted a detailed unpacking here but this is a logical extension of the work.

Alternatively, existing guidelines could be interpreted in the light of the

---

1 The obvious exception here would be the visual ergonomics issues for which standards on resolution, luminance, etc. can be stated quantitatively. Particular aspects of manipulation might also be quantifiable (see for example, Card *et al.* 1978). However, these are very specific instances of HCI that are not singularly concerned with reading.

framework to ensure contextual issues are addressed (thereby lessening one of the major shortcomings of guidelines – their overgeneralised form). For example, the guideline, 'when displaying text that will not fit on a single screen, then use paging rather than scrolling' (Rivlin *et al.* 1990), if applied rigidly, would lead to some very unusable designs. But if interpreted in the context suggested by the framework, that is, for certain users, doing particular tasks with specific texts, it is unlikely to be followed slavishly (and, ultimately, inappropriately) by a well-meaning designer.[2]

As well as being the most suitable presentation format, the framework is intended to support several uses. First, a designer could use it simply as a checklist to ensure that all important components of the text under design are considered. This guards against the reliance on research findings at one level to ensure good design (for example, just following the advice on visual ergonomics, which concludes that certain fonts, polarity and resolution variables can overcome the reading speed deficit). While that advice might be pertinent and applicable, the framework would suggest that it is but one part of the design problem.

Second, it could be used to guide design by allowing a designer to conceptualise the issues to be dealt with in advance of any specification or prototype. In this sense its advocated use is as an advance organiser enabling the designer to organise his thoughts on the problem and highlight attributes of the specification that need to be considered. As discussed in Chapter 8, such an application could lead to significantly more appropriate first specifications or prototypes, lessening the number of iterations required and thereby reducing the time and costs involved in design.

Third, the framework supports the derivation of predictions about readers' performance with a document. The uses made of the framework in the previous chapter highlight its potential value as a predictive tool for a human factors practitioner, adequately familiar with the research in this area, to predict the type of problems a reader will face using an electronic document. It is the author's view that all of the predictions made were easily derived from the framework through the analysis of the various elements and their manifestation or support in the relevant designs, and that few practitioners would face difficulties deriving similarly accurate predictions in other text/task environments.

Finally, but perhaps most practically, the framework has potential evaluative applications. It could be used to guide expert evaluation of a system under development (that is, a usability assessment) and support troubleshooting for weaknesses in design. This proposed use is not unlike

2   The Rivlin *et al.* guidelines are a prime example of the problems inherent in such advisory formats. While they provide generally useful information to designers, the published set contains at least two erroneous suggestions and several, like the cited example, which sound authoritative but generally fail to allow for important contextual variables which negate their recommendation. For a further review of these guidelines see Dillon (1990).

the first use outlined above except it occurs at a different stage in the design process and is intended to support reasoned examination of the quality of an instantiated design. In this role, one could imagine a designer using the framework to check how the system rated on variables such as image quality, the information model it presents, the type of tasks it will support or manipulations it enables.

What next?

We still have a long way to go before we can come close to designing e-texts that compare favourably with paper for most routine uses, though we are making progress. There is still a lot to be learned about screen ergonomics for electronic text, particularly in some of the areas outlined in the literature review, such as screen size, manipulation facilities, search facilities and icon design. These are all current research themes in the human factors discipline but some specific work aimed at the electronic text domain would be useful as it is not always clear how findings from one area of application (for example, visual search) transfer to another.

Working our way up through the framework we can see that at each level there are many issues to study. Even though one might have thought that manipulation was a straightforward issue, the emergence of handheld devices with small screens and styli has brought the issue of screen size and page turning or scrolling back to the fore.

Of course, at the information model level we are now entering the realm of some exciting research possibilities. Digital genres offer the possibilities of new presentation forms, once structured more directly on the cognitive patterning of users and authors, with dynamic reconfigurations for target users possible in real time.

Specifying the design process for e-texts

At the end of a book such as this it is justifiable to ask if the work could be summarised into short, applicable advice for designers. While it is the intention of the author that the framework should fulfill this requirement, a more explicit statement of how electronic text should be designed in accordance with user-centred principles is probably required to extend the framework for its audience. This final section provides a design sequence that involves the framework's components and offers, on the basis of this author's experience, a good chance of successful goal attainment.

Designing a usable electronic document or library therefore involves the following stages:

- Stakeholder identification and user analysis.
- Scenario generation for contexts of use.
- Simulated usage analysis of the text(s) or document(s) involved according to three (at least but not exclusive) criteria:

  How it is used
  Why it is used
  What readers perceive the information to be.

- Investigations of the extent to which the document structure is fixed by existing readers' models.
- Determining the electronic structure by considering the readers' existing models, potential models and the tasks being performed.
- Designing the manipulation facilities required for basic use and ensuring that readers can at least perform these activities simply and intuitively with the mechanisms provided.
- Add value to the system by offering facilities to perform desirable or advantageous activities that are impossible, difficult or time-consuming with paper or previous designs.
- Ensuring image quality is high.
- Testing the system on users performing real tasks and redesigning accordingly.

Box 10.1 Stages in Designing Usable Electronic Text

The first two steps are important and will provide information of direct relevance to the next three steps. The last step is probably the most important although it is often seen as a luxury that cannot be afforded. Failure to test the design is bound to lead to problems as no theoretical models or formal guidelines exist that can even approach the quality of information obtained from observing real users interacting with a system. This is worth repeating:

> no theoretical models or formal guidelines exist that can even approach the quality of information obtained from observing real users interacting with a system.

This applies to the descriptive framework proposed in this book as much as to any existing model in the user literature. These steps represent a complete process and while they will not guarantee success, they offer better prospects of achieving it than many others.

General conclusion

In 1908, Edmund Huey wrote that to understand reading would be the acme of the psychologist's achievements. Now, almost a century later, that statement is perhaps seen to be more accurate with each successive generation of research on the topic. The subtlety and complexity of the reading process makes it a taxing problem for anyone intent on examining it.

The present work has carved out only part of the reading process as its subject matter. In so doing it has drawn on the ideas and themes of several disciplines concerned with reading. While it might have appeared critical, particularly of the work in cognitive psychology, it cannot claim to have explained the process satisfactorily, or to have solved any of the thorny issues of what humans do when they read texts. However, it has led to a perspective; one that aims at improving the quality of the reading process and ensuring that technology does not make us read despite itself, but actively supports us in this quintessential human activity. The question is not 'Should documents be paper or electronic?' but 'How can any presentation medium best satisfy an information need?' There is no simple answer but the analysis of user experience can help us better understand the question in order to best seek the answer.

The continuing prospects for electronic text

Groucho Marx once said of television that we call it a medium because nothing is ever well done. In many ways it seems as if the role of human factors studies of electronic text has been to adopt a similarly Marxist perspective in highlighting problems with the technology, to show that paper is inherently better or to criticise designers and advocates of the electronic medium for overlooking the human issues. It is hoped that this book has not presented a singularly negative view of electronic text but a realistic one, tempered with the optimism that comes from the author's belief that good design is both possible and beneficial. This section, the final one of the book, examines the prospects for electronic text in the light of the work reported.

Twenty years have passed since Jonassen (1982) uttered the memorable (and now punishable by quotation!) phrase: 'in a decade or so, the book as we know it will be as obsolete as is movable type today' (p. 379).

Whatever the facts about movable type in 1982, the book as we know it is certainly far from obsolete in the early twenty-first century. But Jonassen is not alone; the advent of the Web means that his point of view is considered standard and even unquestionable in some quarters, where the truth of this claim lies not in its timescale but in its implications.

The implications of widespread electronic text 'any year now' are important. As this work has attempted to highlight, textual information is everywhere: at home in the form of anything from instructions for operating microwave ovens to the novels that induce sleep; at work in the form of texts ranging from reports on latest developments in company sales to the memos and emails that descend from above; and in the world at large in the form of newspapers, advertising boards, shop catalogues, websites and so on. Avoiding text in contemporary industrial societies would be a feat of Herculean proportions. Modifying text, therefore, by presentation in electronic rather than paper forms, will have a significant impact on our lives.

In this light the zeal of advocates is understandable, it's just that when humans enter into the equation, accurately predicting these impacts becomes difficult, if not impossible. Paper is familiar, is well liked, easy to use (most of the time), affords a representation of its structure that is quickly acquired by readers and leads to the emergence of conventional forms, is portable, supports excellent image quality and is cheap, since publishers have long since recovered their capital outlay on production equipment. Obviously, examples could be found of paper documents which flout such conventional benefits but they hold true for the majority of paper texts, while currently the opposite tends to hold for electronic ones.

The progress of electronic text will be neither explosive nor all-embracing. It will only progress by gaining footholds in small task and text domains and by being found usable there (and possibly, at first, only by a few enthusiasts in these domains). As technology develops, screens improve, portability increases and resistance is lowered the scope for electronic text will broaden, but there is little reason to believe paper will become obsolete in the near future (if ever).

The process will be accelerated by good design, of the kind advocated here, but, conversely, it will be hampered by weak design, that is, design which fails to consider all elements of the framework. It is unlikely that there is anything inherently constraining in the concept of electronic text that cannot be solved by technological improvements and increased knowledge of human information usage. However, the process of reading is not simple and texts are used in multiple ways for myriad tasks by millions of people. In 1996 the Norwegian Ministry of the Environment produced a report indicating that paper consumption had trebled in the previous 30 years, with a further doubling expected by 2010 (OECD 1996). It should be seen as cautionary that the most successful applications of information technology to date have been the word processor and the photocopier.

In the 1994 edition I wrote the following:

> As I write, the computing world is full of claims and predictions for the latest gadget, this one being the Apple Newton and equivalent personal technologies. One interesting prediction is that in a few short years we shall all be reading full-length texts with these wonders of mini-engineering and it strikes me on reading such foolish claims, that some technocrats just never learn. Perhaps the only reasonable prediction that can be made is that we shall witness the emergence of dual-form documents: electronic versions for some tasks, paper versions for others. The strengths of the computer will enable cheap storage and rapid access while the intimacy and familiarity of paper will be retained for detailed studying and examination of material.

Replace the reference to the Newton (now where did that technological innovation go?) with the latest book reader or PDA and not much has changed. I point this out not as a signal of my abilities to foretell the future but to illustrate how well a critical reading of the empirical literature can guide us. A text without a reader is worthless. Similarly, a technology without a user is pointless. The human is the key; only by relating technologies to the needs and capabilities of the user can worthwhile systems be developed. The work in this book is a step in that direction for electronic texts, but there remains a long journey ahead.

Appendix

Example protocol for reader in validity experiment

Time	Comment	Action
0.00		Reads question 1
0.11	I'm going to the Index to see if there's anything on taste	Scrolls
0.17	No . . . Contents	Reads Contents
0.24	No . . . I've a feeling Introduction covers the taste of wine . . . I'll check that	Scrolls down to Introduction
0.31		Scrolls further and reads
0.45	. . . about the colour?	
0.48		Scrolls further and reads
0.53		Scrolls further and reads
0.58		Scrolls further and reads
1.03		Scrolls further and reads
1.06		Has reached new section
1.08		Still reading
1.17	Right I think sweetness is one . . .	Writes down 'Sweetness'
1.20		Scrolls on and reads
1.31		Reaches new section
1.41	I think it's Sweetness and Body, just these two . . .	Scanning text, writes down 'Body'
1.52	I'll just check to see if there's anything later on . . .	Scrolling and reading further
1.55		Reaches new section
1.58		Reaches new section
2.03	No . . . I don't think so . . . I'm going to leave that question . . .	
2.08		Reads question 2

Time	Comment	Action
2.18	I've got a feeling I've just seen that when I was looking for . . . fermentation	Scrolls directly back to top of document and reads the Contents
2.23	Basically the yeast dies . . .	Scrolls to relevant section and scans text
2.37		Confirms answer and writes it down
3.04		Reads question 3
3.10	Something to do with Fermentation? . . .	Scans text, then scrolls down
3.19	Yes . . . to keep the yeast alive and stop the wine burning . . .	Reads text
3.21		Writes down answer
3.41		Reads question 4
3.47	Again, I think I've seen something on that	Scrolls continually down while scanning very quickly
3.55		Scrolls further down
4.00		Still scrolling and scanning rapidly
4.07		Still scrolling, has moved into previously unread text
4.11	Ah . . . I must have passed it . . .	
4.19		Scrolls back through the text scanning rapidly
4.28		Has scrolled back as far as Fermentation section
4.32		Has scrolled back to Introduction
4.37		Reading Contents
4.44	There's nothing in the Contents that's telling me . . . but I'm sure it must be near the start of the document . . . It's fundamental . . .	Reading the early part of the Introduction
4.53	Something to do with colour maybe . . .	Still reading Introduction and following sections Scrolling down as necessary
5.10	But I don't know what you mean by 'lighter' (a reference to the question) or 'taste'	Scrolls further down the Introduction
5.23	. . . Ah . . . it's the caramelisation of residual sugar	Quotes from the text having found a suitable answer

Time	Comment	Action
5.32		Writes down answer
5.51		Reads question 5
6.07	To the Index then . . . I haven't seen anything on this	Scrolls directly down to the bottom of the document
6.11		Scrolls slowly back up to the top of the Index
6.17		Scrolls quickly to the body of the Index
6.19	Grapes on page 1? . . . No . . .	Reading Index terms
6.25		Scrolls directly up to the top of the file and then scrolls slower down to a section in the Introduction
6.56		Starts scrolling back through the Introduction
7.01	It must be in the body of the report then . . .	Reading section on Fermentation again
7.03		Scrolls down to Aging section
7.06		Scrolling and reading the following sections
7.21		Studying the text intensely
7.27		Reading sections on Sweetness and Body
		Scrolling slowly as required
7.31		Reading section on Wine categories: Table and Dessert wines
7.33	Oh . . . Dessert wines	Writes down answer
7.47		Reads question 6
7.51	I've just passed a section on Aging	Scrolls up to Aging
7.55		Reads through section
8.09	Mentions a bit about vintage port . . . doesn't say how old it should be though . . .	
8.22	I think I'll find the section on Port	Goes straight up to Contents
8.25		Browsing through Contents
8.29	No . . . Index	Drags scroll bar down to end
8.38		Drags scroll bar to top

Time	Comment	Action
8.43		Selects Goto command from the menu. Inputs Goto Page 4
8.45		Views sections on Table and Dessert wines
8.51		Scrolls down to section on Aperitifs and Fortified wines
8.53	Port . . .	Finds relevant reference
8.59	Vintage port . . . at least 20 years old	Writes down answer
9.14		Reads question 7
9.17	Haven't seen anything on this method before . . . Solera . . . Check the Index	Drags scroll bar down to end
9.23	What a crappy Index	
9.36		Drags scroll bar back up to middle of text
9.42		Reads serially through the text from section on Aging to section on Sparkling Wines, using slow scroll as necessary
10.30		Drags scroll bar to top to see Contents
10.34	I've just remembered . . . I can search for . . .	Invokes search facilities
10.36		Inputs 'Solera'
10.52		Finds the appropriate answer
10.54	Oh . . . I missed that . . . I skimmed past it	Writes down answer
11.03		Reads question 8
11.07		Drags scroll bar to top to read Contents
11.09	My god . . . I'll search for that again	Invokes search facilities
11.11		Inputs 'Woodworm'
11.21	Continue from beginning . . . ? Yes	Hits return
11.25		End of document message. Search is unsuccessful
11.27		Reads question again
11.35	Wormwood . . . bloody hell	Corrects search term
11.55		Starts search
12.01	Vermouth eh . . .	Term is found in relevant section

Time	Comment	Action
12.03		Writes down answer
12.10		Reads question 9
12.12	That's got something to do with Champagne wines	There's a reference to Sparkling Wines at present position in text
		Reads this.
12.24		Scrolls down text
		Continues reading
12.29	Produces natural effervescence	Writes down answer
12.51		Reads question 10
12.54		Scrolls directly to top of text to see Contents
12.57		Scrolls slowly through contents while reading it
13.03		Invokes Goto facility
		Inputs '10'
13.07		Reading section on Rhône
13.20	Mainly red wines are there . . .	Reads through to Loire section
13.25	Oh yeah . . . mainly white . . . a bit of luck	Writes down answer
13.38		Reads question 11
13.44		Scrolls directly to top to view Contents
13.50	Countries start on page 6	
13.54		Invokes Goto command
		Inputs '6'
14.01		Reading section on France
14.04	France, fairly obviously . . .	Reads through relevant sections on countries
14.24	France and the US, that is in the Italy section	Writes down answer
14.39		Reads question 12
14.42	Types of wine? . . .	Scrolls slowly back through the text
15.00		Scrolls slowly back reading text as he does so
15.47		Reaches Contents
15.52	Ah . . . Aging on page 2?	Scrolls back down
15.59		Reads section on Aging

Time	Comment	Action
16.18	OK . . . it must be those . . .	Makes notes
16.55	I think that must be the reference to it but it's not the same as 'as soon as it's bottled'	Refers to question
17.10	Champagne . . . if you naturally ferment it, it's just going to stay in the bottle . . .	
17.20		Jumps down to Index
17.24		Scrolls slowly back through the Index
17.30	What about Beaujolais	
17.42		Invokes Goto facility and inputs '11'
17.53		Reads sections on Burgundy and Beaujolais
18.20		Scrolls back on section just read
18.24	I'll look for bottled or something	Invokes Search facilities and inputs 'bottle'
18.31	No . . .	Cancels this action
18.38		Reads sections on Loire and Burgundy
18.43		Invokes search facilities again and searches on 'bottle'
18.48		Prompted to continue search from beginning of file he cancels
18.52	No . . .	Tries search again using same term
19.06	No . . . this isn't it	Has found numerous references to 'bottle' in section on Aging
19.13		Cancels Find command
19.19		Invokes search facilities again and searches on same term
19.21		Reads the section on Port and Sherry where search facilities have taken him
19.35		Find Next
19.37	Germany and Italy?	Find Next

Time	Comment	Action
19.41		In Champagne section
		Find Next
19.45		Is in California section
		Find Next then takes him to Sparkling Wines section
19.50		Find Next returns the 'start from beginning message'. He cancels the Find command
19.54	So it's just champagne and semisweet, but that doesn't seem quite right	Writes down answer Session ends

Bibliography

Andre, T., Williges, R. and Hartson, H. (1999) The effectiveness of usability evaluation methods: determining the appropriate criteria. *Proceedings of the Human Factors and Ergonomics Society 43rd Annual Meeting*, pp. 1090–1094. Santa Monica, CA: Human Factors and Ergonomics Society.

Annett, J. and Duncan, K. D. (1967) Task analysis and training design. *Occupational Psychology* 41: 211–221.

Archer, L. (1965) *Systematic Method for Designers*. London: The Design Council.

Asimow, M. (1962) *Introduction to Design*. Englewood Cliffs, NJ: Prentice-Hall.

Ausubel, D. P. (1968) *Educational Psychology: A Cognitive View*. New York: Holt, Rinehart & Winston.

Bailey, R. (1993) Performance versus preference. *Proceedings of the Human Factors Society Annual Conference*, pp. 282–286. Santa Monica, CA: Human Factors and Ergonomics Society.

Bailey, W. A., Knox, S. T. and Lynch, E. F. (1988) Effects of interface design upon user productivity. *Proceedings of CHI*, 88: pp. 207–212. New York: ACM.

Baker, N. (2001) *Double Fold: Libraries and the Assault on Paper*. New York: Random House.

Bannister, D. and Mair, D. (1968) *The Evaluation of Personal Constructs*. London: Academic Press.

Barber, P. (1988) In favour of theory. *Ergonomics* 31(6): 871–872.

Bartlett, F. C. (1932) *Remembering*. Cambridge: Cambridge University Press.

Bauer, D. and Cavonius, C. R. (1980) Improving the legibility of visual display units through contrast reversal, in E. Grandjean and E. Vigliani (eds), *Ergonomic Aspects of Visual Display Terminals*. London: Taylor & Francis.

Bauer, D., Bonacker, M. and Cavonius, C. R. (1983) Frame repetition rate for flicker-free viewing of bright VDU screens. *Displays* January: 31–33.

Bazerman, C. (1988) *Shaping Written Knowledge*. Wisconsin: University of Madison Press.

Becker, D. and Dwyer, M. (1994) Using hypermedia to provide learner control. *Journal of Educational Multimedia and Hypermedia* 3(2): 155–172.

Beldie, I. P., Pastoor, S. and Schwartz, E. (1983) Fixed versus variable letter width for televised text. *Human Factors* 25(3): 273–277.

Belmore, S. (1985) Reading computer presented text. *Bulletin of the Psychonomic Society* 23(1): 12–14.

Beyer, H. and Holtzblatt, K. (1998) *Contextual Design*. San Francisco: Morgan Kaufmann.

Binder, A. (1964) Statistical theory, in P. Farnsworth, O. McNemar and Q. McNemar (eds), *Annual Review of Psychology*, vol. 15, pp. 277–310.

Bishop, A. (1998) Digital libraries and knowledge disaggregation: the use of journal article components. Proceedings of ACM Conference on Digital Libraries 1998, New York: ACM Press, 29–38.

Boehm, (1988) A spiral model of software development and enhancement. IEEE Computer, 21, 2, 61–72.

Borgman, C. (2000) *From Guttenberg to the Global Information Infrastructure. Access to Information in the Networked World*. Boston: MIT Press.

Boyarski, D., Neuwirth, C., Forlizzi, J. and Regli, S. (1998) A study of fonts designed for screen display, in C.-M. Karat (ed.), *CHI '98: Human Factors in Computing Systems*, pp. 87–94. Reading, MA: Addison-Wesley, ACM Press.

Brand, J. and Judd, K. (1993) Angle of hard copy and text editing performance. *Human Factors* 35(1): 57–69.

Brewer, W. (1987) Schemas versus mental models in human memory, in I. P. Morris (ed.), *Modelling Cognition*, pp. 187–197. London: John Wiley & Sons.

Buckingham, B. (1931) New data on the typography of textbooks. *Yearbook of the National Society for the Study of Education* 30: 93–125.

Buckley, P. (1989) Expressing research findings to have a practical influence on design, in J. Long and A. Whitefield (eds), *Cognitive Ergonomics and Human Computer Interaction*, pp. 166–190. Cambridge: Cambridge University Press.

Buie, E. and Hassell, E. (1982) *An Introduction to Wines*. Electronic document delivered with HyperTIES Software.

Bush, V. (1945) As we may think. *Atlantic Monthly* 176(1): 101–108.

Byrne, M., John, B., Wehrle, N. and Crow, D. (1999) The tangled web we wove: a taskonomy of WWW user, in *CHI'99, Proceedings of the ACM SIGCHI Conference*, pp. 544–551. New York: ACM Press.

Cakir, A., Hart, D. J. and Stewart, T. F. M. (1980) *Visual Display Terminals*. Chichester: John Wiley & Sons.

Calvi, L. (1997) Navigation and disorientation: a case study. *Journal of Educational Multimedia and Hypermedia* 6(3/4): 305–320.

Card, S., English, W. and Burr, B. (1978) Evaluation of mouse, rate-controlled isometric joystick, step keys and text keys for text selection on a CRT. *Ergonomics* 21: 601–613.

Card, S. K., Mackinlay, J. and Shneiderman, B. (eds) (1999) *Readings in Information Visualization; Using Vision to Think*. San Francisco: Morgan Kaufmann.

Card, S. K., Moran, T. P. and Newell, A. (1983) *The Psychology of Human–Computer Interaction*. Hillsdale, NJ: Lawrence Erlbaum Associates.

Carroll, J. and Campbell, R. (1986) Softening up hard science: a reply to Newell and Card. *Human–Computer Interaction* 2(3): 227–249.

Chalmers, A. (1982) *What is This Thing Called Science?* Milton Keynes: Open University Press.

Chapanis, A. (1988) Some generalizations about generalizations. *Human Factors* 30(3): 253–268.

Chapman, L. J. and Hoffman, M. (1977) *Developing Fluent Reading*. Milton Keynes: Open University Press.

Charney, D. (1987) Comprehending non-linear text: the role of discourse cues, in *Proceedings of Hypertext '87*, pp. 109–120. Chapel Hill: University of North Carolina.

Chatman, S. (1990) *Coming to Terms: The Rhetoric of Narrative in Fiction and Film*. Ithaca, NY: Cornell University Press.

Chen, C. and Czerwinski, M. (1997) Spatial ability and visual navigation: an empirical study. *New Review of Hypermedia and Multimedia* 3: 67–90.

Clark, D. (1983) Reconsidering research on learning from media. *Review of Educational Research* 53(4): 445–459.

Cohen, G. (1988) *Memory in the Real World*. London: Lawrence Erlbaum Associates.

Collier, G. (1987) Thoth-II: hypertext with explicit semantics. *Proceedings of Hypertext '87*, pp. 269–289. Chapel Hill: University of North Carolina.

Conklin, J. (1987) Hypertext: an introduction and survey. *IEEE Computer* 20(9): 17–41.

Creed, A., Dennis, I. and Newstead, S. (1987) Proof-reading on VDUs. *Behaviour and Information Technology* 6(1): 3–13.

Cross, N. (1985) Styles of learning, designing and computing. *Design Studies* 3: 157–162.

Crowder, R. and Wagner, R. (1992) *The Psychology of Reading: An Introduction*, 2nd edn. New York: Oxford University Press.

Cuff, R. (1980) On casual users. *International Journal of Man–Machine Studies* 12: 163–187.

Curtis, B. (1990) Empirical studies of the software design process, in D. Diaper, D. Gilmore, G. Cockton and B. Shackel (eds), *INTERACT'90*, pp. xxxv–xl. North Holland: Amsterdam.

Cushman, W. H. (1986) Reading from microfiche, VDT and the printed page: subjective fatigue and performance. *Human Factors* 28(1): 63–73.

Darke, J. (1979) The primary generator and the design process. *Design Studies* 3: 157–162.

de Beaugrande, R. (1980) *Text, Discourse and Process*. Norwood, NJ: Ablex.

de Beaugrande, R. (1981) Design criteria for process models of reading. *Reading Research Quarterly* 16(2): 261–315.

De Brujin, D., De Mul, S. and Van Oostendorp, H. (1992) The influence of screen size and text layout on the study of text. *Behaviour and Information Technology* 11: 71–78.

Diaper, D. (1990) Simulation: stepping stones between requirements and design, in M. Life, C. Narborough-Hall and W. Hamilton (eds), *Simulation and the User Interface*, pp. 59–72. London: Taylor & Francis.

Dillon, A. (1988) The role of usability labs in systems design, in T. Megaw (ed.), *Contemporary Ergonomics*, pp. 63–73. London: Taylor & Francis.

Dillon, A. (1990) A review of Rivlin *et al.* (1990): guidelines for screen design. *Hypermedia* 2(2): 171–173.

Dillon, A. (1991) Requirements analysis for hypertext applications: the why, what and how approach. *Applied Ergonomics* 22(4): 458–462.

Dillon, A. (1992) Reading from paper versus screens: a critical review of the empirical literature. *Ergonomics* 35(10): 1297–1326.

Dillon, A. (1994) *Designing Usable Electronic Text*, 1st edn. London: Taylor & Francis.

Dillon, A. (2001) Beyond usability: process, outcome and affect in human–computer interactions. *Canadian Journal of Library and Information Science* 26(4): 57–69.

Dillon, A. (2002) Information Architecture in SASIST: just where did we come from? *Journal of the American Society for Information Science and Technology* 53(10): 821–823.

Dillon, A. and Gabbard, R. (1998) Hypermedia as an educational technology: a review of the empirical literature on learner comprehension, control and style. *Review of Educational Research* 68(3): 322–349.

Dillon, A. and Gushrowski, B. (2000) Genres and the Web – is the home page the first digital genre? *Journal of the American Society for Information Science* 51(2): 202–205.

Dillon, A. and McKnight, C. (1990) Towards a classification of text types: a repertory grid approach. *International Journal of Man–Machine Studies* 33(6) 623–636.

Dillon, A. and Morris, M. (1996) User acceptance of new information technology – theories and models, in M. Williams (ed.), *Annual Review of Information Science and Technology*, vol. 31, pp. 3–32. Medford, NJ: Information Today.

Dillon, A. and Schaap, D. (1996) Expertise and the perception of structure in discourse. *Journal of the American Society for Information Science* 47(10): 786–788.

Dillon, A. and Vaughan, M. (1997) 'It's the journey and the destination': shape and the emergent property of genre in evaluating digital documents. *New Review of Hypermedia and Multimedia* 3: 91–94.

Dillon, A. and Watson, C. (1996) User analysis HCI – the historical lessons from individual differences research. *International Journal of Human–Computer Studies* 45(6): 619–638.

Dillon, A., Richardson, J. and McKnight, C. (1990) Navigation in hypertext: a critical review of the concept, in D. Diaper, D. Gilmore, G. Cockton and B. Shackel (eds), *INTERACT '90*, pp. 587–592. North Holland: Amsterdam.

Dillon, A., Richardson, J. and McKnight, C. (1991) Institutionalising human factors in the design process: the ADONIS experience. *Contemporary Ergonomics '91*, pp. 421–426. London: Taylor & Francis.

Dillon, A., Sweeney, M. and Maguire, M. (1993) A survey of usability evaluation practices and requirements in the European IT industry, in J. Alty, S. Guest and D. Diaper (eds), *HCI '93. People and Computers VII*. Cambridge: Cambridge University Press.

Drury, C. G., Paramore, B., Van Cott, H. P., Grey, S. M. and Corlett, E. N. (1988) Task analysis, in G. Salvendy (ed.), *Handbook of Human Factors*. New York: John Wiley & Sons.

Dryden, L. M. (1994) Literature, student-centered classrooms, and hypermedia environments, in C. Selfe and S. Hilligoss (eds), *Literacy and Computers: The Complications of Teaching and Learning with Technology*, pp. 282–304. New York: The Modern Language Association of America.

Duchnicky, R. L. and Kolers, P. A. (1983) Readability of text scrolled on a visual display terminal as a function of window size. *Human Factors* 25(6): 683–692.

Duffy, T., Palmer, J. and Mehlenbacher, B. (1992) *On-Line Help: Design and Evaluation*. Norwood, NJ: Ablex.

Dyson, M. and Haselgrove, M. (2001) The influence of reading speed and line length on the effectiveness of reading from screen. *International Journal of Human–Computer Studies* 54: 585–612.

Dyson, M. and Kipping, G. (1998) The effect of line length and method of movement on patterns of reading from screen. *Visible Language* 32: 150–181.

Eason, K. (1988) *Information Technology and Organisational Change.* London: Taylor & Francis.

Eason, K. (2001) Changing perspectives on the organizational consequences of information technology. *Behaviour and Information Technology* 20(5): 323–328.

Edwards, D. and Hardman, L. (1989) 'Lost in Hyperspace': cognitive mapping and navigation in a hypertext environment, in R. McAleese (ed.), *Hypertext: Theory into Practice*, pp. 105–125. Oxford: Intellect.

Egan, D., Remde, J., Landauer, T., Lochbaum, C. and Gomez, L. (1989) Behavioural evaluation and analysis of a hypertext browser. *Proceedings of CHI '89*, pp. 205–210. New York: Association of Computing Machinery.

Elkerton, J. and Williges, R. (1984) Information retrieval strategies in a file search environment. *Human Factors* 26(2): 171–184.

Ellis, A. (1983) *Reading, Writing and Dyslexia.* London: Lawrence Erlbaum Associates.

Elm, W. and Woods, D. (1985) Getting lost: a case study in interface design. *Proceedings of the Human Factors Society 29th Annual Meeting*, pp. 927–931. Santa Monica, CA: Human Factors Society.

Engelbart, D. (1963) A conceptual framework for the augmentation of man's intellect, in P. Howerton and D. Weeks (eds), *Vistas in Information Handling*, vol. 1, pp. 1–29. London: Spartan Books.

Ericsson, K. A. and Simon, H. A. (1984) *Protocol Analysis.* Cambridge, MA: MIT Press.

Ewing, J., Mehrabanzad, S., Sheck, S., Ostroff, D. and Shneiderman, B. (1986) An experimental comparison of a mouse and arrow-jump keys for an interactive encyclopedia. *International Journal of Man–Machine Studies* 24(1): 29–45.

Feldman, T. (1990) *The Emergence of the Electronic Book.* British National Bibliography Research Fund Report 46, The British Library.

Ferguson, G. (1959) *Statistical Analysis in Psychology and Education.* New York: McGraw-Hill.

Fleishman, E. and Quaintance, M. (1984) *Taxonomies of Human Performance.* Orlando, FL: Academic Press.

Forester, T. (ed.) (1985) *The Information Technology Revolution.* Oxford: Blackwell.

Gardener, A. and McKenzie, J. (1988) *Human Factors Guidelines for the Design of Computer-Based Systems.* London: DTI/MOD. Available from HUSAT Research Institute, Loughborough University of Technology.

Gardiner, M. and Christie, B. (eds) (1987) *Applying Cognitive Psychology to User–Interface Design.* Chichester: John Wiley & Sons.

Garland, J. (1982) Ken Garland and associates: designers – 20 years work and play. Cited in R. Waller (1987) The typographic contribution to language: towards a model of typographic genres and their underlying structures. PhD thesis, Department of Typography and Graphic Communication, University of Reading.

Garnham, A. (2001) *Mental Models and the Interpretation of Anaphora.* Hove, East Sussex: Psychology Press.

Gerrig, R. and McKoon, G. (1998) The readiness is all; the functionality of memory-based text processing. *Discourse Processes* 26(2): 67–86.

Gould, J. D. and Grischkowsky, N. (1984) Doing the same work with hard copy and cathode-ray tube (CRT) computer terminals. *Human Factors* 26(3): 323–337.

Gould, J. D., Alfaro, L., Barnes, V., Finn, R., Grischkowsky, N. and Minuto, A. (1987a) Reading is slower from CRT displays than from paper: attempts to isolate a single variable explanation. *Human Factors* 29(3): 269–299.

Gould, J. D., Alfaro, L., Finn, R., Haupt, B. and Minuto, A. (1987b) Reading from CRT displays can be as fast as reading from paper. *Human Factors* 29(5): 497–517.

Gould, J., Boies, S. and Ukelson, J. (1997) How to design usable systems, in M. Helander, T. Landauer and P. Prabhu (eds), *Handbook of Human Computer Interaction*, pp. 231–254. Amsterdam: Elsevier.

Gray, W. D., John, B. E. and Atwood, M. E. (1993) Project Ernestine: validating a GOMS analysis for predicting and explaining real-world task performance. *Human–Computer Interaction* 8(3): 237–249.

Hammond, N. and Allinson, L. (1987) The travel metaphor as design principle and training aid for navigating around complex systems, in D. Diaper and R. Winder (eds), *People and Computers III*, pp. 75–90. Cambridge: Cambridge University Press.

Hammond, N. and Allinson, L. (1989) Extending hypertext for learning: an investigation of access and guidance tools, in A. Sutcliffe and L. Macaulay (eds), *People and Computers V*, pp. 293–304. Cambridge: Cambridge University Press.

Hammond, N., Gardiner, M., Christie, B. and Marshall, C. (1987) The role of cognitive psychology in user–interface design, in M. Gardiner and B. Christie (eds), *Applying Cognitive Psychology to User–Interface Design*, pp. 13–52. Chichester: John Wiley & Sons.

Harrison, M. and Thimbleby, H. (eds) (1990) *Formal Methods in Human Computer Interaction*. Cambridge: Cambridge University Press.

Hartley, J. (1985) *Designing Instructional Text*, 2nd edn. London: Kogan Page.

Hassard, J. (1988) FOCUS as a phenomenological technique for job analysis: its use in multiple paradigm research. *International Journal of Man–Machine Studies* 27: 413–433.

Hatt, F. (1976) *The Reading Process*. London: Clive Bingley.

Helander, M. G., Billingsley, P. A. and Schurick, J. M. (1984) An evaluation of human factors research on visual display terminals in the workplace, in F. Muckler (ed.), *Human Factors Review*, pp. 55–129. Santa Monica, CA: Human Factors Society.

Horney, M. (1993) A measure of hypertext linearity. *Journal of Educational Multimedia and Hypermedia* 2(1): 67–82.

Hudson, L. (1968) *Frames of Mind*. Harmondsworth: Penguin.

Huey, E. B. (1908) *The Psychology and Pedagogy of Reading*. New York: Macmillan.

Jaschinski-Kruza, W. (1990) On the preferred viewing distances to screen and document at VDU workplaces. *Ergonomics* 33(8): 1055–1063.

John, B. and Marks, S. (1997) Tracking the effectiveness of usability evaluation methods. *Behaviour and Information Technology* 16(4/5): 188–202.

John, B. and Vera, A. (1992) A GOMS analysis of a graphic, machine paced, highly interactive task. *Proceedings of CHI'92*, pp. 251–258, New York: ACM Press.

Johnson-Laird, P. (1983) *Mental Models*. Cambridge: Cambridge University Press.

Jonassen, D. (1982) *The Technology of Text. Vol. I. Principles for Structuring, Designing, and Displaying Text*. Englewood Cliffs, NJ: Educational Technology Publications.

Jones, S. (2002) The internet goes to college: how students are living in the future with today's technology. http://www.pewinternet.org/reports/index.asp. Downloaded December 12, 2002.

Julien, H. and Duggan, L. (2000) A longitudinal analysis of the information needs and uses literature. *Library and Information Science Research* 22: 1–19.

Just, M. A. and Carpenter, P. (1980) A theory of reading: from eye movements to comprehension. *Psychological Review* 87(4): 329–354.

Kak, A. V. (1981) Relationships between readability of printed and CRT-displayed text. *Proceedings of the Human Factors Society – 25th Annual Meeting*, pp. 137–140. Santa Monica, CA: Human Factors Society.

Kelly, D. and Cool, C. (2002) The effects of topic familiarity on information search behavior, in *Proceedings of the Second ACM/IEEE Joint Conference on Digital Libraries*, pp. 74–75. New York: ACM Press.

Kelly, G. (1955) *The Psychology of Personal Constructs* (2 vols). New York: Norton.

Kerlinger, F. (1986) *Foundations of Behavioral Research*. New York: Holt, Rinehart & Winston.

Kerr, S. T. (1986) Learning to use electronic text: an agenda for research on typography, graphics, and interpanel navigation. *Information Design Journal* 4(3): 206–211.

Kieras, D. and Polson, P. (1985) An approach to the formal analysis of user complexity. *International Journal of Man–Machine Studies* 22: 365–394.

Kintsch, W. (1974) *The Representation of Meaning in Memory*. Hillsdale, NJ: Lawrence Erlbaum Associates.

Kintsch, W. and Yarborough, J. (1982) The role of rhetorical structure in text comprehension. *Journal of Educational Psychology* 74: 828–834.

Klein, L. and Eason, K. (1991) *Putting Social Science to Work*. Cambridge: Cambridge University Press.

Kline, P. (1988) *Psychology Exposed, or The Emperor's New Clothes*. London: Routledge.

Kolers, P. A., Duchnicky, R. L. and Ferguson, D. C. (1981) Eye movement measurement of readability of CRT displays. *Human Factors* 23(5): 517–527.

Landauer, T. (1987) Relations between cognitive psychology and computer systems design, in J. Carroll (ed.), *Interfacing Thought*, pp. 1–25. Cambridge, MA: MIT Press.

Landauer, T. (1995) *The Trouble with Computers*. Cambridge, MA: MIT Press.

Landow, G. (1992) *Hypertext: The Convergence of Contemporary Critical Theory and Technology*. Baltimore: Johns Hopkins University Press.

Lawson, B. (1979) *How Designers Think*. London: Architectural Press.

Lee, E., Whalen, T., McEwen, S. and Latrémouille, S. (1984) Optimizing the design of menu pages for information retrieval. *Ergonomics* 27(10): 1051–1069.

Lehto, M., Zhu, W. and Carpenter, B. (1995) The relative effectiveness of hypertext and text. *International Journal of Human Computer Interaction* 7(4): 293–313.

Lewis, C. (1994) Sample sizes for usability studies: additional considerations. *Human Factors* 36: 368–378.

Lewis, C. and Wharton, C. (1997) Cognitive walkthroughs, in M. Helander, T. Landauer and P. Prabhu (eds), *Handbook of Human–Computer Interaction*, pp. 717–732. Amsterdam: North Holland.

Leventhal, L., Teasley, B., Instone, K., Rohlman, D. and Farhat, J. (1993) Sleuthing in HyperHolmes: an evaluation of using hypertext versus a book to answer questions. *Behaviour and Information Technology* 12(3): 149–164.

Licklider, J. (1965) *Libraries of the Future*. Cambridge, MA: MIT Press.

Lovelace, E. A. and Southall, S. D. (1983) Memory for words in prose and their locations on the page. *Memory and Cognition* 11(5): 429–434.

Lund, A. M. (1997) Expert ratings of usability maxims. *Ergonomics in Design*. 5(3): 15–20.

Macdonald, S. and Stevenson, R. (1996) Disorientation in hypertext: the effects of three text structures on navigation performance. *Applied Ergonomics* 27: 61–68.

Mandler, J. and Johnson, N. (1977) Remembrance of things parsed: story structure and recall. *Cognitive Psychology* 9: 111–151.

Martin, A. (1972) A new keyboard layout. *Applied Ergonomics* 3(1): 42–51.

Martin, L. and Platt, M. (2001) Printing and screen reading in the medical school curriculum: Guttenberg vs. the cathode ray tube. *Behaviour and Information Technology* 20(3): 143–148.

Mayes, D., Sims, V. and Koonce, J. (2001) Comprehension and workload differences for VDT and paper-based reading. *International Journal of Industrial Ergonomics* 28: 367–378.

McAleese, R. (1989) Navigation and browsing in hypertext, in R. McAleese (ed.), *Hypertext: Theory into Practice*, pp. 6–44. Oxford: Intellect.

McClelland, J. L. and Rumelhart, D. (1981) An interactive activation model of context effects in letter perception: Part 1. An account of basic findings. *Psychological Review* 88: 375–407.

McConkie, G., Underwood, N., Zola, D. and Wolverton, G. (1985) Some temporal characteristics of processing during reading. *Journal of Experimental Psychology: Human Perception and Performance* 11(2): 168–186.

McKnight, C., Richardson, J. and Dillon, A. (1989) The authoring of hypertext documents, in R. McAleese (ed.), *Hypertext: Theory into Practice*, pp. 138–147. Oxford: Intellect.

McKnight, C. (2000) The personal construction of information space. *Journal of the American Society for Information Science* 51(8): 730–733.

McKnight, C., Dillon, A. and Richardson, J. (1990) A comparison of linear and hypertext formats in information retrieval, in R. McAleese and C. Green (eds), *Hypertext: State of the Art*, pp. 10–19. Oxford: Intellect.

McKnight, C., Dillon, A. and Richardson, J. (1991) *Hypertext in Context*. Cambridge: Cambridge University Press.

McKnight, C., Dillon, A., Richardson, J., Haraldsson, H. and Spinks, R. (1992) Information access in different media: an experimental comparison, in E. J. Lovesey (ed.), *Contemporary Ergonomics*, pp. 515–519. London: Taylor & Francis.

Miller, R. B. (1967) Task taxonomy: science or technology? In W. Singleton, R. Easterby and D. Whitfield (eds), *The Human Operator in Complex Systems*. London: Taylor & Francis.

Mills, C. B. and Weldon, L. J. (1986) Reading text from computer screens. *ACM Computing Surveys* 19(4) 329–358.

Milner, N. (1988) A review of human performance and preference with different input devices to computer systems, in D. Jones and R. Winder (eds), *People and Computers IV*, pp. 341–362. Cambridge: Cambridge University Press.

Mitchell, D. (1982) *The Process of Reading: A Cognitive Analysis of Fluent Reading and Learning to Read*. New York: John Wiley & Sons.

Moran, T. (1981) The command language grammar, a representation for the user interface of interactive computer systems. *International Journal of Man–Machine Studies* 15(1): 3–50.

Muter, P. and Mauretto, P. (1991) Reading and skimming from computer screens and books: the paperless office revisited? *Behaviour and Information Technology* 10(4): 257–266.

Muter, P., Latremouille, S. A., Treurniet, W. C. and Beam, P. (1982) Extended reading of continuous text on television screens. *Human Factors* 24(5): 501–508.

Neal, A. and Darnell, M. (1984) Text editing performance with partial line, partial page and full page displays. *Human Factors* 26(4): 431–441.

Nelson, T. (1987) *Literary Machines* (Abridged electronic version 87.1). San Antonio.

Newell, A. and Card, S. (1985) The prospects for psychological science in human–computer interaction. *Human–Computer Interaction* 1: 209–242.

Nielsen, J. (1990) *Hypertext and Hypermedia*. New York: Academic Press.

Nielsen, J. (1992) Finding usability problems through heuristic evaluation, in *Proceedings of CHI'92*, pp. 373–380. New York: ACM.

Nielsen, J. (1993) *Usability Engineering*. Cambridge, MA: Academic Press Professional.

Nisbett, R. and Wilson, T. (1977) Telling more than we can know: verbal reports on mental processes. *Psychological Review* 84: 231–259.

Norman, D. (1988) *The Psychology of Everyday Things*. New York: Basic Books.

Norman, D. and Draper, S. (eds) (1986) *User Centred System Design*. Hillsdale, NJ: Lawrence Erlbaum Associates.

Norman, K. (1991) *The Psychology of Menu Navigation*. New York: Ablex.

Norman, K. and Chen, J. (1988) The effect of tree structure on search in a hierarchical menu selection system. *Behaviour and Information Technology* 7(1): 51–65.

Oborne, D. and Holton, D. (1988) Reading from screen versus paper: there is no difference. *International Journal of Man–Machine Studies* 28(1): 1–9.

O'Hara, K. and Sellen, A. (1997) A comparison of reading paper and on-line documents, in *Proceedings of CHI'97*, pp. 335–342. New York: ACM Press.

Oppenheim, A. N. (1966) *Questionnaire Design and Attitude Measurement*. London: Heinemann.

OECD (Organization for Economic Co-operation and Development) (1996) Report for the OECD Workshop on: Rethinking Paper Consumption. OCDE/GD(97)111, Paris.

Pask, G. (1976) Styles and strategies of learning. *British Journal of Education Psychology* 46: 128–148.

Pastoor, S., Schwartz, E. and Beldie, I. P. (1983) The relative suitability of four dot-matrix sizes for text presentation on colour television screens. *Human Factors* 25(3): 265–272.

Pearce, B. (ed.) (1984) *Health Hazards of VDUs?* Chichester: John Wiley & Sons.

Pearson, R. G. and Byars, G. E. (1956) The development and validation of a checklist for measuring subjective fatigue. Report # TR-56-115. San Antonio, TX: USAF School of Aviation Medicine.

Pettigrew, K. and McKecknie, L. (2001) the use of theory in information science research. *Journal of the American Society for Information Science and Technology* 52(1): 62–73.

Piolat, A., Roussey, J.-Y. and Thunin, O. (1997) Effects of screen presentation on screen reading and revising. *International Journal of Human–Computer Studies* 47: 565–589.

Pollock, A. and Hockley, A. (1997) What's wrong with internet searching. *D-Lib Magazine.* http://www.dlib.org/dlib/march97/bt/03pollock.html.

Popper, K. (1986) *The Poverty of Historicism.* Reading: ARK (originally published in 1957).

Pugh, A. (1975) The development of silent reading, in W. Latham (ed.), *The Road to Effective Reading.* London: Ward Lock.

Radl, G. W. (1980) Experimental investigations for optimal presentation mode and colours of symbols on the CRT screen, in E. Grandjean and E. Vigliani (eds), *Ergonomic Aspects of Visual Display Terminals*, pp. 127–137. London: Taylor & Francis.

Rasmussen, J. (1986) *Information Processing and Human–Machine Interaction: an Approach to Cognitive Engineering.* London: North Holland.

Reiffel, E. (1999) The genre of mathematics writing and its implications for digital documents, in *Proceedings of the 32nd Annual Hawaii International Conference on System Sciences.* Los Alamitos, CA: IEEE Computer Society (published on CD-ROM).

Rho, Y. and Gedeon, T. (2000) Reading patterns and formats of academic articles on the web. *SIGCHI Bulletin* 32(1): 67–72.

Richardson, J., Dillon, A., McKnight, C. and Saadat-Samardi, M. (1988) The manipulation of screen presented text: experimental investigation of an interface incorporating a movement grammar. HUSAT Memo #431, Loughborough University of Technology.

Richardson, J., Dillon, A. and McKnight, C. (1989) The effect of window size on reading and manipulating electronic text, in E. Megaw (ed.), *Contemporary Ergonomics*, pp. 474–479. London: Taylor & Francis.

Rivlin, C., Lewis, R. and Cooper, R. (1990) *Guidelines for Screen Design.* Oxford: Blackwell Scientific.

Rosson, M. and Carroll, J. (2001) *Usability Engineering.* San Francisco: Morgan Kaufmann.

Rothkopf, E. Z. (1971) Incidental memory for location of information in text. *Journal of Verbal Learning and Verbal Behavior* 10: 608–613.

Rumelhart, D. (1977) Toward an interactive model of reading, in S. Dornic (ed.), *Attention and Performance VI.* Hillsdale, NJ: Erlbaum.

Ryle, A. (1976) Some clinical applications of grid technique, in P. Slater (ed.), *The Measurement of Intrapersonal Space by Grid Technique* (2 vols). London: John Wiley & Sons.

Samuels, S. and Kamil, M. (1984) Models of the reading process, in P. Pearson. (ed.), *Handbook of Reading Research*, pp. 185–224. New York: Longman.

Sauter, S., Gottlieb, M., Rohrer, K. and Dodson, V. (1983) The well-being of video display terminal users: an exploratory study. Report no. 210-79-0034. Cincinnati, OH: US Department of Health and Human Sciences.

Schumacher, G. and Waller, R. (1985) Testing design alternatives: a comparison of procedures, in T. Duffy and R. Waller (eds), *Designing Usable Texts*, pp. 377–403. Orlando, FL: Academic Press.

Schwartz, E., Beldie, I. and Pastoor, S. (1983) A comparison of paging and scrolling for changing screen contents by inexperienced users. *Human Factors* 25: 279–282.

Shackel, B. (1959) Ergonomics for a computer. *Design* 120: 36–39.

Shackel, B. (1986) Ergonomics in design and usability, in M. Harrison and A. Monk (eds), *People and Computers: Designing for Usability*. Cambridge: Cambridge University Press.

Shackel, B. (1991) Usability – context, framework, definition, design and evaluation, in B. Shackel and S. Richardson (eds), *Human Factors for Informatics Usability*, pp. 21–37. Cambridge: Cambridge University Press.

Shaw, M. L. G. (1980) *On Becoming a Personal Scientist*. London: Academic Press.

Shaw, M. L. G. and Gaines, B. (1987) KITTEN: knowledge initiation and transfer tools for experts and novices. *International Journal of Man–Machine Studies* 27: 251–280.

Sheih, K. (2000) Effects of reflection and polarity on LCD viewing distance. *International Journal of Industrial Ergonomics* 35(3): 275–282.

Shneiderman, B. (1997) *Designing the User Interface: Strategies for Effective Human–Computer Interaction*. San Francisco: Addison Wesley.

Simpson, A. (1989) Navigation in hypertext: design issues. Paper presented at International OnLine Conference '89, London.

Simpson, A. (1990) Towards the design of an electronic journal. Unpublished PhD thesis, Department of Human Sciences, Loughborough University of Technology.

Simpson, A. and McKnight, C. (1990) Navigation in hypertext: structural cues and mental maps, in R. McAleese and C. Green (eds), *Hypertext: State of the Art*, pp. 73–83. Oxford: Intellect.

Slater, P. (1976) *The Measurement of Intrapersonal Space by Grid Technique* (2 vols). London: John Wiley & Sons.

Smedshammar, H., Frenckner, K., Nordquist, C. and Romberger, S. (1989) Why is the difference in reading speed when reading from VDUs and from paper bigger for fast readers than for slow readers? Paper presented at WWSU 1989, Second International Conference, Montreal.

Smith, A. and Savory, M. (1989) Effects and after-effects of working at a VDU, in E. Megaw (ed.), *Contemporary Ergonomics*. London: Taylor & Francis.

Smith, F. (1978) *Reading*. Cambridge: Cambridge University Press.

Smoliar, S. and Baker, J. (1997) Text types in hypermedia, in *Proceedings of the 30th Annual Hawaii International Conference on Systems Science*, vol. V1, 68–77. Hawaii: Wailea.

Sommerville, I. (1997) *Software Engineering*, 5th edn. New York: Addison Wesley.

Spinks, R. (1991) An experimental comparison of hypertext and paper for information location in lengthy text. Unpublished MSc thesis, Department of Human Sciences, Loughborough University of Technology.

Spring, M. (1991) *Electronic Printing and Publishing: The Document Processing Revolution*. New York: Marcel Dekker, Inc.

Stark, H. (1990) A comparison of jump and pop-up windows in hypertext, in R. McAleese and C. Green (eds), *Hypertext: State of the Art*, pp. 2–9. Oxford: Intellect.

Starr, S. J. (1984) Effects of video display terminals in a business office. *Human Factors* 26(3): 347–356.

Stevens, G. C. (1983) User-friendly computer systems?: a critical examination of the concept. *Behaviour and Information Technology* 2(1): 3–16.

Suchman, L. (1988) *Plans and Situated Action*. Cambridge: Cambridge University Press.

Tinker, M. A. (1958) Recent studies of eye movements in reading. *Psychological Bulletin* 55: 215–231.

Tinker, M. A. (1963) *Legibility of Print*. Ames, Iowa: Iowa State University Press.

Tombaugh, J., Lickorish, A. and Wright, P. (1987) Multi-window displays for readers of lengthy texts. *International Journal of Man–Machine Studies* 26: 597–616.

Tovey, M. (1986) Designing with both halves of the brain. *Design Studies* 5(4): 219–228.

Tractinsky, N. (1997) Aesthetics and apparent usability: empirically assessing cultural and methodological issues. *Proceedings of the Annual ACM SIGCHI Conference: CHI '97*, pp. 115–121. New York: ACM Press.

Trigg, R. and Suchman, L. (1989) Collaborative writing in NoteCards, in R. McAleese (ed.), *Hypertext: Theory into Practice*, pp. 45–61. Norwood, NJ: Ablex.

Tyrrell, R. A., Pasquale, T. B., Aten, T. and Francis E. L. (2001) Empirical evaluation of user responses to reading text rendering using ClearType™ technology. *SID 01 DIGEST*, p. 1205.

Underwood, G. and Batt, V. (1996) *Reading and Understanding*. Oxford: Blackwell.

van Dijk, T. A. (1980) *Macrostructures*. Hillsdale, NJ: Lawrence Erlbaum Associates.

van Dijk, T. A. and Kintsch, W. (1983) *Strategies of Discourse Comprehension*. London: Academic Press.

Vaughan, M. (1999) Identifying regularities in users' conceptions of information spaces: designing for structural genre convention and mental representations of structure for web-based newspapers. Unpublished PhD dissertation, Indiana University.

Vaughan, M. and Dillon, A. (2000) Learning the shape of information – a longitudinal study. *Proceedings of the ACM DL00 Conference on Digital Libraries*, pp. 236–237. New York: ACM Press.

Venezky, R. L. (1984) The history of reading research, in P. Pearson (ed.), *Handbook of Reading Research*, pp. 3–38. New York: Longman.

Waller, R. (1987) The typographic contribution to language: towards a model of typographic genres and their underlying structures. Unpublished PhD thesis, Department of Typography and Graphic Communication, University of Reading.

Weinberg, G. (1971) *The Psychology of Computer Programming*. New York: Van Nostrand Reinhold.

Whitefield, A. (1989) Constructing appropriate models of computer users: the case of engineering designers, in J. Long and A. Whitefield (eds), *Cognitive*

Ergonomics and Human Computer Interaction, pp. 66–94. Cambridge: Cambridge University Press.

Wickens, C. (1991) *Engineering Psychology and Human Performance*. Columbus: Charles Merrill.

Wilkinson , R. T. and Robinshaw, H. M. (1987) Proof-reading: VDU and paper text compared for speed, accuracy and fatigue. *Behaviour and Information Technology* 6(2): 125–133.

Witkin, H. A., Moore, C. A., Goodenough, D. R. and Cox, P.W. (1977) Field dependent and field independent cognitive styles and their educational implications. *Review of Educational Research* 47(1): 1–64.

Wittgenstein, L. (1953) *Philosophical Investigations*. New York: Macmillan.

Wright, P. (1980) Textual literacy: an outline sketch of psychological research on reading and writing, in P. Kolers, M. Wrolstad and H. Bouma (eds), *Processing of Visible Language* 2, pp. 517–535. London: Plenum Press.

Wright, P. (1998) Designing information supported performance: the scope for graphics. *Journal of Computer Documentation* 22: 3–15.

Wright, P. and Lickorish, A. (1983) Proof-reading texts on screen and paper. *Behaviour and Information Technology* 2(3): 227–235.

Wright, P. and Lickorish, A. (1988) Colour cues as location aids in lengthy texts on screen and paper. *Behaviour and Information Technology* 7(1): 11–30.

Wright, P. and Lickorish, A. (1990) An empirical comparison of two navigation systems for two hypertexts, in R. McAleese and C. Green (eds), *Hypertext: State of the Art*, pp. 84–95. Oxford: Intellect.

Zechmeister, E., McKillip, J., Pasko, S. and Bespalec, D. (1975) Visual memory for place on the page. *Journal of General Psychology* 92: 43–52.

Index

Printed and bound by CPI Group (UK) Ltd, Croydon, CR0 4YY

01/11/2024

01782632-0016